74 **Topics in Current Chemistry**

Fortschritte der Chemischen Forschung

Organic Compounds

Syntheses / Stereochemistry / Reactivity

Springer-Verlag
Berlin Heidelberg GmbH 1978

This series presents critical reviews of the present position and future trends in modern chemical research. It is addressed to all research and industrial chemists who wish to keep abreast of advances in their subject.

As a rule, contributions are specially commissioned. The editors and publishers will, however, always be pleased to receive suggestions and supplementary information. Papers are accepted for "Topics in Current Chemistry" in English.

ISBN 978-3-662-15824-1   ISBN 978-3-540-35945-6 (eBook)
DOI 10.1007/978-3-540-35945-6

Library of Congress Cataloging in Publication Data. Main entry under title: Organic compounds. (Topics in current chemistry ; 74) Bibliography: p. Includes index. 1. Chemistry, Physical organic--Addresses, essays, lectures. I. Vögtle, Fritz, 1947- II. Lewis, Edward S., 1920- III. Series. QD1.F58 vol. 74 [QD476] 540'.8s [547] 78-3465

2152/3140−543210

Contents

Stereochemistry of Multibridged, Multilayered, and Multistepped Aromatic Compounds – Transanular Steric and Electronic Effects

Fritz Vögtle and Gerd Hohner

Institut für Organische Chemie und Biochemie der Universität Bonn, Gerhard-Domagk-Str. 1, D-5300 Bonn, Germany

Table of Contents

1. Introduction

[2.2]Paracyclophane (*1*)[1, 2] is characterized by the close vicinity of the phenylene rings, which are yoked face to face in parallel planes. Their distance is substantially (by more than 0.3 Å) shorter than the van der Waals distance in common crystalline arenes. This strange topology gives rise to the consequences indicated below:

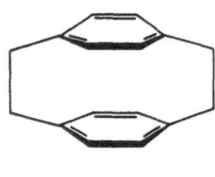

1

    a) The aromatic π-electrons show significant *transanular interaction,* thus affecting especially the absorption and emission spectra. This effect stimulated numerous theoretical discussions[3–10]. Also, the reactivity of the compound towards electrophilic reagents is markedly influenced by such electronic through space effects[11].

    b) The extraordinary geometry of compound *1* causes a serious *out-of-plane-deformation of the aromatic rings;* X-ray data show that they have a boat shape[12]. Hence, strained molecules of the cyclophane type contribute to the current topic[13, 14]: how bent can aromatic systems be without loosing their distinctive properties (diatropy, regenerative chemical behaviour etc.)?

    Further approaching and deforming the benzene rings by additional short bridges ought to enhance both of the peculiarities of [2.2]paracyclophane. The recently synthesized highly strained, *multiply bridged* phane hydrocarbons (*e.g. 2, 15, 16, 19, 21*) thus turn out to be novel subjects of considerable theoretical interest. Their often very consuming syntheses made borderlands of preparative and theoretical organic chemistry accessible.

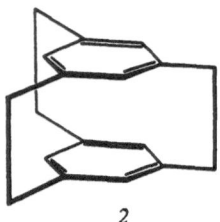

2

    Besides such energy-rich, sterically overcrowded cyclophanes, their *homologues,* connected by longer bridges, are worth considering as well. They allow to correlate transanular interaction to the distance separating the interacting parts of the molecule. This distance can be controlled by varying sort and number of the bridging atoms. The more easily accessible *hetero-multibridged* phanes (*e.g. 37–43*) often serve as starting materials for preparing strained phane hydrocarbons *via* ring contraction (extrusion) reactions.

In recent years further novel classes of compounds were added to cyclophane chemistry. The *multilayered phanes*[15] derived from [2.2]paracyclophane contain coaxially stacked benzene rings connected by ethano bridges *para* to one another. The first members of this series were described in 1964[16]. Quadruple layered phane hydrocarbons *3* and *4* reveal in their UV-spectra long range electronic effects penetrating several arene units.

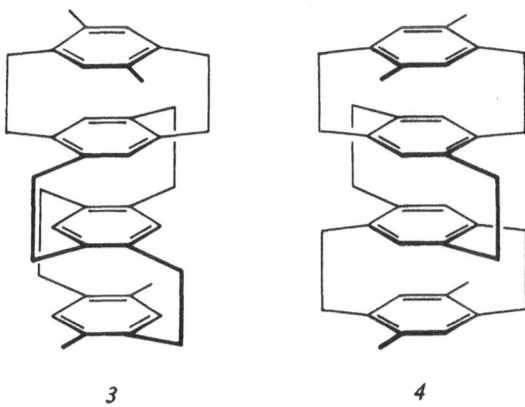

3   4

Analogously, repeating the [2.2]metacyclophane building principle leads to *multistepped phanes, e.g. 5*[17]. The possibility of constructing alternative (configurationally isomeric) structures by either *up-* or *down*-linkage of the neighbouring benzene rings opens up new attractive stereochemical aspects (see below).

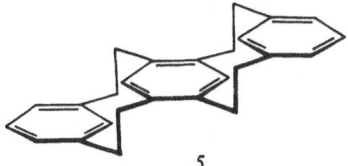

5

2. Multibridged Aromatic Compounds

2.1. [2.2.2](1,3,5)Cyclophane (*2*) and [2.2.2](1,3,5)Cyclophane-1,9,17-triene (*6*)

The synthesis and properties of the highly symmetrical hydrocarbon phanes *2* and *6* have been reported in 1970[18]. X-ray data of *6*[19] indicate a molecular strain exceeding that of [2.2]paracyclophane-diene (*7*). The latter was estimated by Gantzel and Trueblood to be about 39 kcal/mole (163.0 kJ/mole)[20].

Figure 1 shows the geometry of the triene *6*[19], Fig. 2, for comparison, that of the diene *7*[21]. In the case of the [2.2.2]phane the three bridges force a chairshape deformation upon the benzene rings, which show a minimum distance of 2.76 Å (C3–C15, C5–C11, C7–C13). In contrast, the average van der Waals distance be-

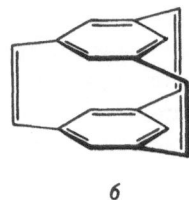

6

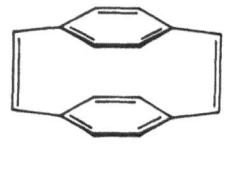

7

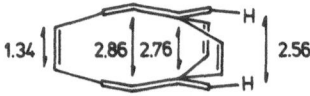

Fig. 1. Geometry of cyclophanetriene 6[19]

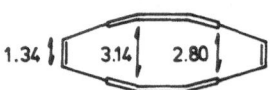

Fig. 2. Geometry of cyclophanediene 7[21]

tween parallel arene units was found to be 3.4 Å[22]. The nonplanar arrangement is expected to rehybridize the aromatic carbon atoms, giving them more sp^3-character[a].

Rehybridization would partially displace the intraanular π-electron density to the outside of the benzene rings, thus reducing destabilizing interaction in the interior of the molecule. In addition, rehybridization rationalizes a striking feature of the triene 6, also found in [2.2]paracyclophane 1[12], its diene 7[21] and in [3.3]paracyclophane[20]: the aromatic C—H-bonds of all phane hydrocarbons mentioned are inclined towards the inside of the molecule, e.g. in triene 6 by an angle of 13°.

Besides ring deformation and repulsive π-π-interaction, the considerable angle deformation of the C_{aryl}—$C_{methyne}$-bonds also contributes to the overall strain of molecule 6. These bonds are bent out of the plane defined by the three adjacent aromatic carbon atoms by an angle of c. 24°[19]. Despite the mutual repulsion of the benzene rings, the olefinic bonds are of normal length[b]. Due to its high symmetry, the [2.2.2]phane 2 shows only two singlet signals in the ^{1}H—NMR-spectrum. The resonance of the aromatic H-atoms appears at unusually high field (δ = 5.73 ppm). High field shift is a characteristic feature of face-to-face phanes[27], and was attributed to two factors pushing into the same direction[27b]:

a) the magnetic anisotropy effect of the opposite arene ring

b) the rehybridization of the aromatic carbon atoms due to ring distortion. The hence arising disturbance of the ring current was expected to diminish deshielding of the arene protons.

However, already Longone and Chow pointed out that rehybridization should play a minor role in diamagnetic shift[16]. This in indicated by the ^{1}H—NMR-spectra of the more strained members of the [m]paracyclophane series 8. Despite even a drastic ring distortion their phenylene protons resonate in the usual range (see Table 1). Triene 6 displays two singlet signals of equal intensity (δ = 6.24 and 7.37 ppm). By

---

a) However, the ^{13}C—NMR-spectrum does not support the rehybridization concept. The coupling constant value $^1J_{CH}$ (162 MHz) falls into the range typical of undisturbed arenes[23].

b) This also holds for all [2.2]cyclophane dienes so far measured by X-ray analysis[21, 24] and clearly contrasts with the lengthening of the CH_2—CH_2-bonds in [2.2]phanes bridged by aliphatic chains[12, 25, 26].

comparison with partially deuteriated *6* the high field signal was assigned to the arene protons[18].

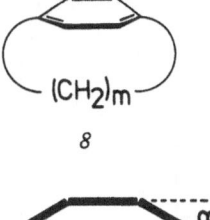

8

Fig. 3. Geometry of [m]paracyclophanes

Table 1. [1]H−NMR-data and ring deformation of some [m]paracyclophanes

	δH-arene (ppm)	α
[8]Paracyclophane	7.15 (m)[1][28]	c. 9°[1][28]
[7]Paracyclophane	7.07[29]	c. 17°[2][30]
[6]Paracyclophane	7.17[31]	22.4° (calc.)[32]

[1]) 4-Carboxylic acid.
[2]) 3-Carboxylic acid.

The UV-spectrum of *2* shows two transitions at 258 nm (ϵ = 1200) and 312 nm (ϵ = 96)[23]. Triene *6* having presumably more distorted but wider separated benzene rings than *2*[c] has bands at 252 nm (ϵ = 1960) and 325 nm (ϵ = 90). The long wavelength maximum of [2.2]paracyclophane *1* appears at 302 nm (ϵ = 155)[3a]. The bathochromic shift of the latter with respect to the corresponding absorptions of open-chained, strainfree reference compounds, was attributed to the concomitant effects of ring deformation and transanular π-π-interactions in accord with theoretical estimations[3c]. In [2.2]paracyclophane, the orbitals of the aromatic π-electrons are oriented parallelly to the bridging CH_2-CH_2-bonds, thus, according to Gleiter[9], giving rise to a band in the range of 280 to 300 nm caused by π-π-interaction.

The same effects which are responsible for the spectroscopic anomalies of system *1* should also play an important role in the absorptional behaviour of the [2.2.2]phanes *2* and *6*. It still appears difficult, however, to arrive at a definite correlation of the individual contributing factors with observed data[33].

Treating *2* with $CH_3COCl/AlCl_3$ yields the monoacetyl derivative *9* by electrophilic substitution[23]. The deformed benzene rings, thus, act in a regenerative ("aromatic") manner with respect to this reaction. The extremely high reaction rate re-

---

[c]) Compare the analogous situation in the case of compounds *1* and *7*[1].

minds of the similar behaviour of [2.2]paracyclophane and is rationalized analogously[11a]: transanular delocalization of the positive charge stabilizes the transition state preceding the σ-complex.

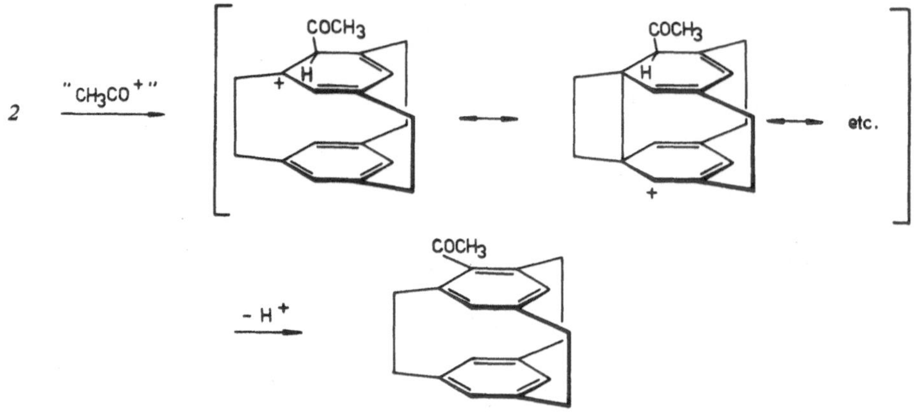

9

In contrast, reaction of 2 with HCl/NaCN/ZnCl$_2$ does not yield the expected formylation product. Instead, a mixture is obtained containing neither aldehydic nor olefinic nor aromatic protons. Treatment with HCl/AlCl$_3$ leads to aliphatic adducts of composition $2 \cdot H_x Cl_{6-x}$, which are reduced by Li/tert-butanol to the novel cage skeleton 10. The reaction, involving repeated transanular bond formation, is believed to proceed via the mechanism outlined below[23].

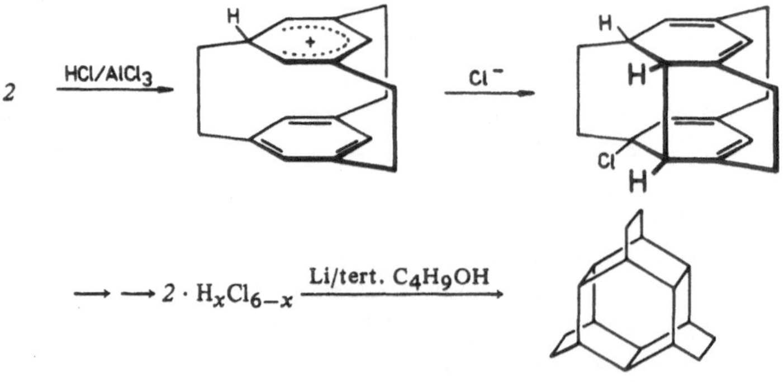

10

2.2. [2.2.2](1,2,4)Cyclophane (11) and [3.2.2](1,2,5)Cyclophane (12)

The ^{1}H–NMR-spectrum of the less symmetric 2-isomer 11 described by Truesdale and Cram[34] is, again, characterized by a distinct upfield-shift of the arene proton signals. Molecular models indicate a nonplanar arrangement of the benzene rings.

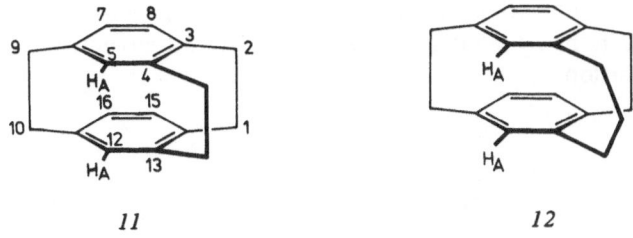

11   *12*

Ring atoms C4 and C13 as well as C5 and C12 come closer together than do C7/C16 and C8/C15. The protons H_A attached to C5 and C12, therefore, submerge deeper into the anisotropic region of the opposite ring and thus resonate at higher field ($\delta = 6.05$ ppm) than the other arene protons ($\delta = 6.17$ and 6.37 ppm). The less strained molecule *12*[34], however, shows the H_A-signal at lowest field. According to UV-data, *12* should closely resemble [2.2]paracyclophane with respect to transanular π-π-interaction and ring deformation. The spectra of both hydrocarbons nearly coincide. *11* displays λ_{max}-values similar to *12*, whereas the extinction coefficients widely differ. In contrast to [2.2]paracyclophane (*1*), *11* forms a (mono) Diels-Alder adduct with dimethyl acetylenedicarboxylate. This points to more "cyclohexatriene" character of the benzene rings in *11* than in *1*. On the other hand, *11* readily undergoes electrophilic acetylation.

2.3. [2.2.2](1,2,4)(1,3,5)Cyclophane (*13*)

The asymmetric hydrocarbon phane *13* reported recently[35] is extremely strained according to the molecular model and is inclined to resinify even at ambient temperature[d]. The unusually high field shifted ^{1}H–NMR-absorption of one arene proton ($\delta = 5.04$ ppm) indicates, in accord with the model, a noncongruent arrangement of the benzene rings.

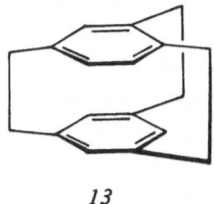

13

2.4. [m.m.m](1,3,5)Cyclophanes (*14*)(m > 2)

Symmetrical triply bridged phane hydrocarbons of type *14* were reported by Hubert and Dale in 1965[36]. Their synthetic method – cyclizing trimerization of appropriate acetylene precursors – made the members *14a–14g* accessible[36, 37], but failed with the highly strained [2.2.2]phane *2*. In its UV-spectrum *14a* differs from its higher homologues chiefly by less intensive absorption and by quenched vibrational

---

[d] However, *13* was prepared by sulfone pyrolysis at 520 °C!

fine structure[37]. From comparison with the homologues of the [m.n]paracyclophane series[3b], it is concluded that in the *14*-series only *14a* shows relevant transanular π-π-interaction and ring distortion.

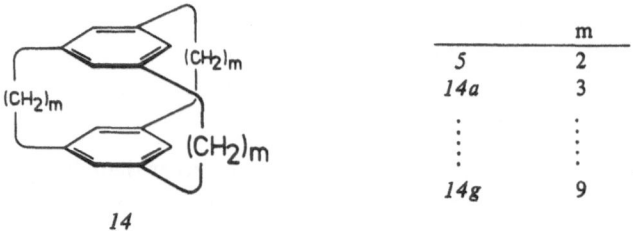

	m
5	2
14a	3
⋮	⋮
14g	9

Cycles *14* form colored molecule ("charge transfer") complexes with tetracyanoethylene (TCNE)[37]. Again, *14a*, which forms an adduct displaying an extraordinary long wave shifted, intense absorption band, drops of line. Similar studies, concerning the complexing behaviour of [m.n]paracyclophanes, resulted in scaling the intramolecular π-π-interactions by the λ_{max}-values of the TCNE-adducts[38]. It was assumed that the bathochromic shift is mostly determined by the extent to which transanular charge transfer occurs from the uncomplexed to the complexed ring (see Fig. 4).

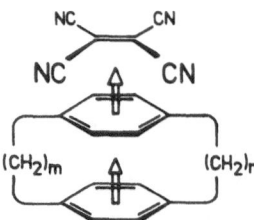

Fig. 4. Assumed charge transfer-stabilization in [m.n]paracyclophane/TCNE-complexes

In [3.3.3]phane *14a*, the transfer of negative charge density called forth by π-π-interaction ought to be favored relative to the higher homologues because of the short distance between the benzene rings. Thus, the exceptional behaviour of the *14a*/TCNE-adduct would be easily interpreted. However, as complexing experiments recently uncovered, the "TCNE method" does not seem to be suitable for determining π-π-interactions in cyclophane molecules[39]. In accord with earlier results of Dewar and Thompson[40], the λ_{max}-values of arene/TCNE-adducts were found to be governed mainly not by charge transfer forces but by field and inductive effects depending intricately on the structure of the complexed arene. However, there is clear evidence of transanular π-π-interaction at least in the case of [2.2]paracyclophane[39].

2.5. [2.2.2.2](1,2,4,5)Cyclophane (*15*) and [2.2.2.2](1,2,3,5)Cyclophane (*16*)

The fourfold bridged phane hydrocarbons *15*[41] and *16*[42] should well represent the closest face-to-face approach of two benzene rings yet performed. However, the arene

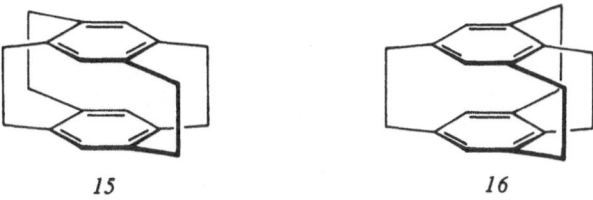

15   16

protons of *15* absorb at lower field strength (δ = 5.96 ppm) than those of [2.2.2]-phane *2* (δ = 5.73 ppm). The quadruply substituted benzene rings in *15* are supposed to be boatshaped, with the tertiary carbon atoms thrusted outwards, thus less subjecting the corresponding protons to the shielding influence of the opposite ring than in molecule *2*. The UV-spectrum of *15* shows moderate bathochromic shifts with respect to [2.2]paracyclophane *1*[3c)] in the middle and short wave length region, whereas the long wave length bands have nearly identical λ_{max}-values (*1:* λ_{max} = 302 nm; *15:* λ_{max} = 303 nm). As to spectroscopic properties *16* strikingly resembles its isomer *15*. The arene protons also resonate at δ = 5.96 ppm, the positions of the UV-absorption maxima show only little difference. Nothing is yet known about the chemical properties of the cycles *15* and *16*.

2.6. [4.4.4.4.4.4](1,2,3,4,5,6)Cyclophane (*18a*)

The first phane hydrocarbon connecting two benzene rings by six bridges was claimed to be synthesized in 1973 by trimerization of the cyclic diyne *17*[43a)]. Yet the compound previously assigned the "percyclophane-4" structure *18a* later has been shown to be the triyne *18b*[43b)]:

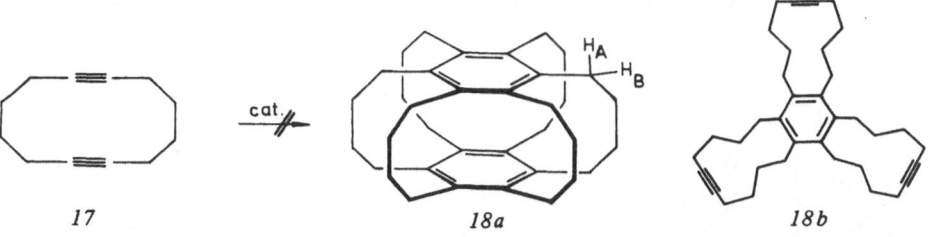

17   18a   18b

Sixfold bridged cage compounds such as *18a* therefore remain unattained synthetic targets.

2.7. Phane Hydrocarbons Containing Extended Arene Units

The molecular strain, which is located in a few bonds in [2.2.2](1,3,5)cyclophane (*2*), is distributed over a mediocyclic frame in the phenylogous triphenylbenzeno-phane *19*[44)]. With respect to 1,3,5-tri-*p*-tolylbenzene as a reference, the arene proton signals of compounds *19* and *20* are only shifted by 0.4 to 0.6 ppm towards higher field owing to the wider separation of the molecular halves. Surprisingly, the UV-spectrum of compound *19* hardly differs from that of 1,3,5-tri-*p*-tolylbenzene[44b)].

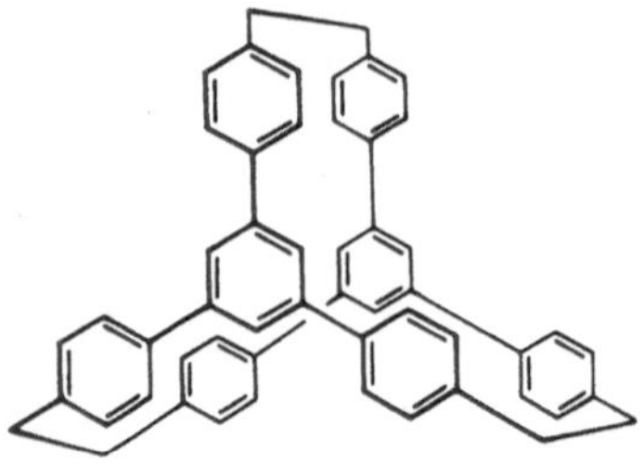

19 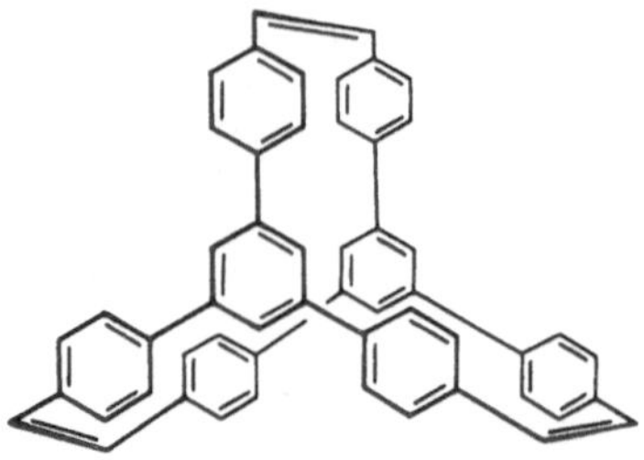   20

It is, however, not clear, in which way π-π-interaction, if there, should particularly affect the triphenylbenzene chromophore. Moreover, a bathochromic effect could be compensated by deformation of the triphenylbenzene units forcibly induced by the molecular geometry of the phane molecule. Relative to *19*, the absorption maximum of *20* is shifted by 8 nm towards longer wave-length. As an explanation the existence of "residual conjugation" between the paraphenylene rings and the olefinic bridges has to be taken into account: at least certain conformations of the flexible molecule ought to allow partial overlap of the formally orthogonal π-systems.

In the phane hydrocarbons *21* and *22*[44], the triphenylmethane units are fixed in a propeller-like, chiral conformation, according to the spectra: the arene protons of each compound produce, instead of an AA'BB' multiplet expected for a flexible molecule, two widely spread AB systems, one appearing at unusually high field (Figs. 5 and 6). These signals could easily be assigned, if one assumes that in both phane molecules opposing phenylene rings are not arranged like mirror-images, but rather are fixed in mutually shifted conformations as shown in Fig. 7.

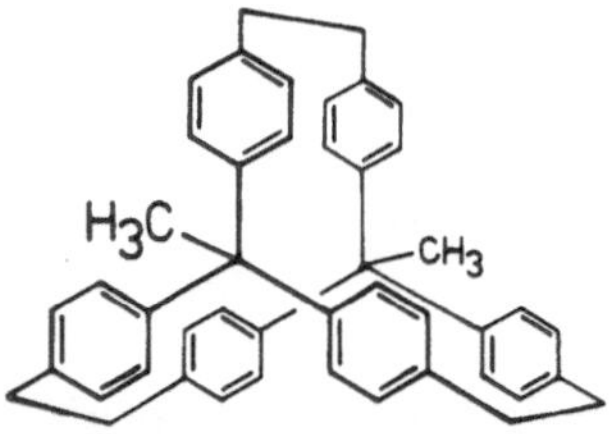

21 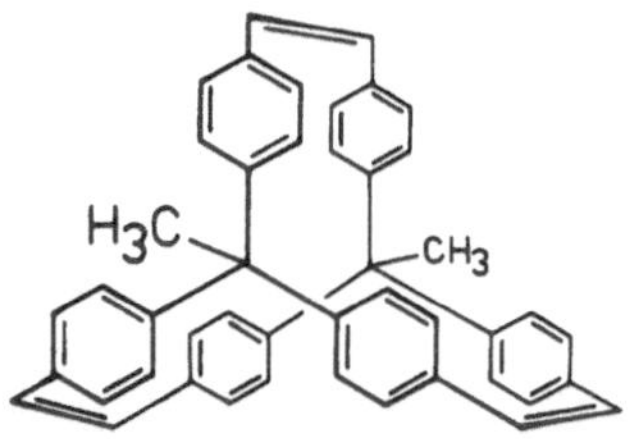   22

Consequently, two of the four protons of each ring (H_A and H_B) come to lie in the shielding anisotropy region of the opposite ring, thus absorbing as a high-field AB-pattern, which is further split by *meta* coupling (*21*: δ = 5.72 ppm; *22*: δ = 5.84 ppm). Protons H_K and H_L do not experience any diatropic magnetic influence and appear in the typical arene proton range [AB-patterns, centered at δ = 7.24 ppm (*21*) resp. δ = 7.29 ppm (*22*)]. In *21*, the rigid conformation not only differentiates H_A and H_B from H_K and H_L, but also the bridge protons H_X from H_Y, creating an AA'BB' spin system centered at δ = 2.82 ppm.

10

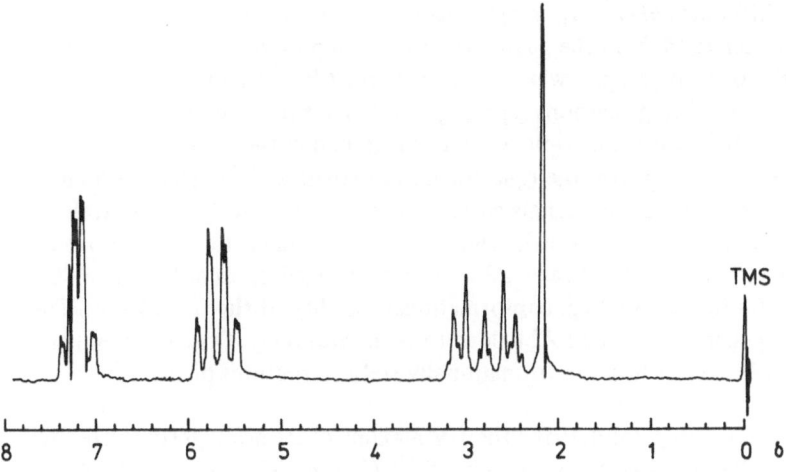

Fig. 5. ¹H–NMR-spectrum of phane hydrocarbon *21* (in CDCl₃)[44b]

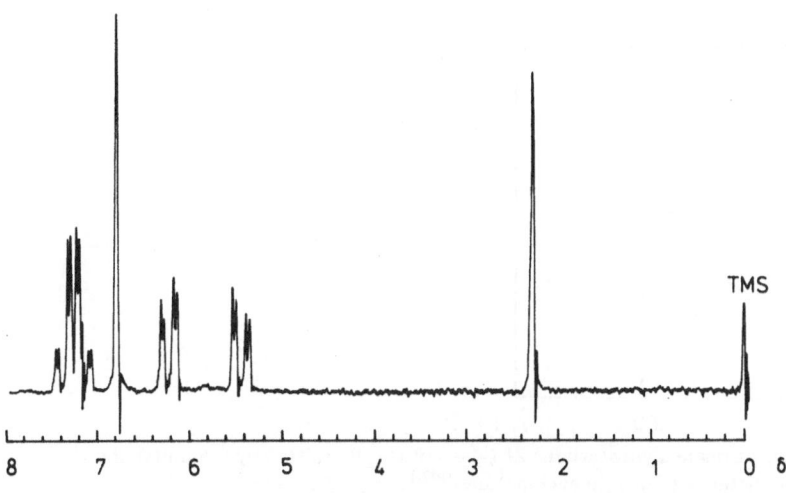

Fig. 6. ¹H–NMR-spectrum of phane hydrocarbon *22* (in CDCl₃)[44b]

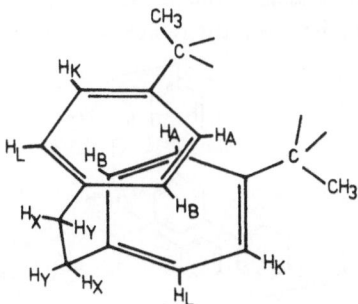

Fig. 7. Assumed geometry of phane hydrocarbon *21* (detail)

11

The shift difference δH_A-δH_B is significantly greater in triene *22* than in *21*. This is easily understood from the geometry of the compounds: the angle enclosed by two opposite benzene rings is wider in *22* than in *21* because of the different hybridizations of the bridge carbons (sp^2 resp. sp^3). Thus protons H_A and H_B experience as more different anisotropic shielding in *22* than in the flatter molecule *21*. ^{13}C–NMR-spectroscopy proves the rigid fixation of triene *22*[44b]. The proton decoupled spectrum displays nine signals corresponding to nine groups of equivalent carbon atoms while a conformationally flexible molecule ought to show seven signals.

The ^{1}H–NMR-spectra of *21* resp. *22* remain unchanged up to at least 175 °C resp. 130 °C. This indicates a high conformational stability of the two phanes. The "molecular propellers"[45] *21* and *22* belong to point group D$_3$. Organic chemistry up to now only knows of four conformationally stable structures of this symmetry[46–49].

Relative to 1,1,1-tri-*p*-tolylethane the UV maxima of *21* are slightly shifted to the red, may be due to transanular interaction[44b] (Fig. 8). The spectrum of *22*, showing higher extinctions and no vibrational structure, indicates partial conjugation of phenylene rings and bridging double bonds.

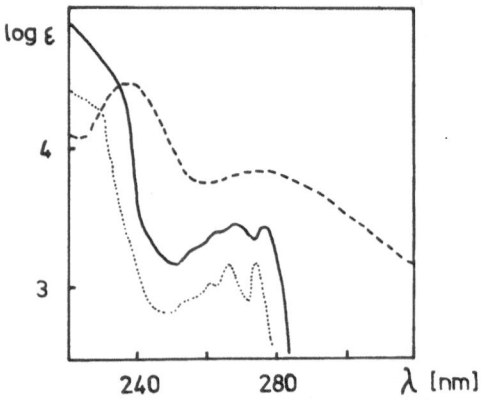

Fig. 8. UV-spectra of phane hydrocarbons *21* (——; vertically shifted by 0.5 units), *22* (————), and 1,1,1-tri-*p*-tolylethane (. . . .) (in cyclohexane)[44b]

The novel diradicaloid hydrocarbon *23*, which may be prepared[50] in a similar manner as *21*, may exist as two triply bridged but electronically independent triphenylmethyl radicals (*23a*). Alternatively, interaction between the radical centers

23a　　　　*23b*　　　　*24*

may result in partial bond formation (*23b*). *23b* shows structural analogies to the hypothetical hexaphenylethane erraneously[51, 52] claimed as triphenylmethyl dimer for a long time[53].

The aminium *24* — the synthesis is under preparation[50] — interests from a similar point of view. The odd electron might skip over to the opposite arene as was demonstrated for [n.n]paracyclophane radical anions (n < 4)[54].

In the ^{1}H—NMR-spectra of the biphenylophanes *25* and *26*[55] the H_C-signals show the greatest, the AA'BB'-multiplets arising from protons H_A and H_B the smallest high field shift, hinting that the doubly bridged benzene rings are situated more closely than the para-phenylene rings.

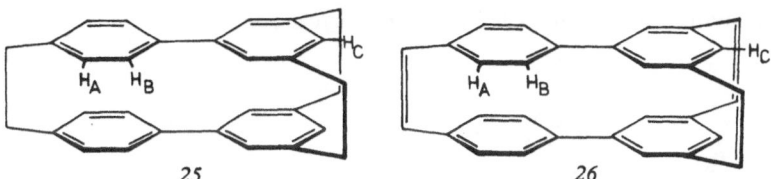

25   26

Since all the arene protons of *26* resonate at lower field than those of *25*, the wider bond angle at sp^2 carbon is assumed to keep the arene units in the olefin at a distance greater than in the CH_2CH_2-bridged phane[e].

The UV-spectra of both biphenylophanes reveal characteristic differences with respect to 3,4',5-trimethylbiphenyl as a reference. The main maxima show a bathochromic shift of about 10 nm; in the long wave length region new maxima or shoulders are observed, most presumably caused by π-π-interactions.

Pyrolysis[56] of tetrasulfone *27* yields a hydrocarbon of composition $C_{32}H_{28}$, which has been given structure *28b* for evident reasons[57]. However, ^{1}H—NMR-data seem to rather favor its diastereomer *28a* containing parallel biphenyl units. Above

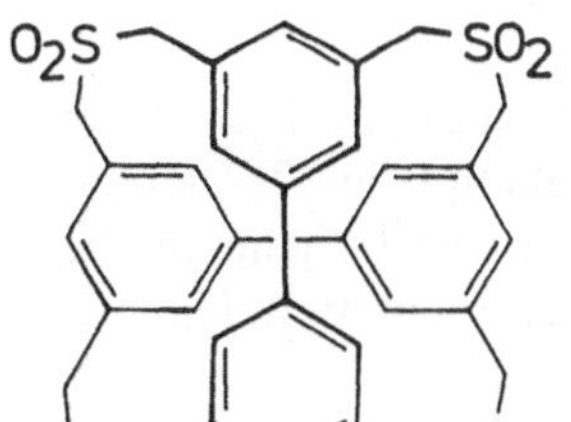

27 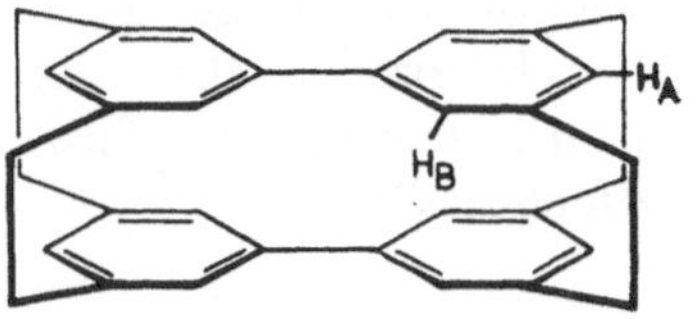   28a 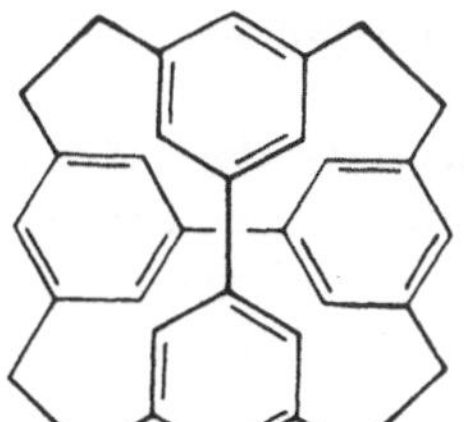   28b

all, the significant highfield shift of the signal assigned to the H_A-protons (δ = 5.93 ppm) supports face-to-face arranged arenes. Strikingly, the H_B-protons resonate in the normal arene proton range (δ = 7.02 ppm). Eventually, the two benzene rings in one biphenyl unit are forced into coplanarity. Thus, the H_B-protons are subjected to the

---

e) The maximum distance of the benzene rings is 3.09 Å in [2.2]paracyclophane (*1*)[12] and 3.14 Å in [2.2]paracyclophane-1,9-diene (*7*)[21]. Compare also the $\delta_{H_{arene}}$-differences in the couples *2/6* (Δδ = 0.51 ppm)[23] and *19/20* (Δδ = 0.1 ppm)[44b].

deshielding influence of the other benzene ring in the very same biphenyl. This influence might counterbalance the shielding effect of the face-to-face ring current. Also, for an explanation, mutual steric compression of the H_B-hydrogen atoms must be taken into account (van der Waals effect)[58]. Definite structural assignment would be accessible only by X-ray analysis.

2.8. Triply Bridged (1,3,5)(1,4)Benzeno<5>phanes

The cyclophane hexaene 29 has been synthesized via a sixfold Wittig reaction[59]. Spectroscopic properties and possible conformations of the hydrocarbon were discussed. Catalytic hydrogenation of 29 led to the polycycle 30.

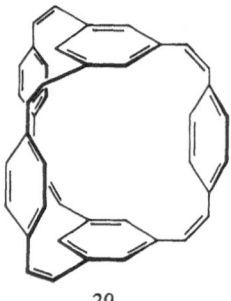
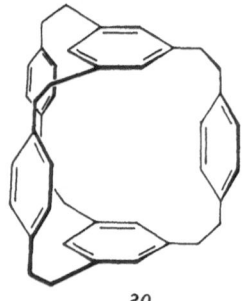

29   30

2.9. "Open Chain Cyclophanes"

The phenylene rings of 1,8-diphenylnaphthalene (31) are arranged in a face-to-face manner[60, 61]. Accordingly, a relatively low δ-value (6.85 ppm[f]) is found for the corresponding protons[61].

Bridging the phenylene rings in the [3.2]paracyclophane analogue 32 reported recently, shifts the phenylene signals by additional c. 0.3 ppm upfield[g][62]. Molecules of the 31 type, as well as, for example, the hydrocarbon 34[64] ("janusene"), or the helicenes[65], all containing arene units in a face-to-face orientation, may be regarded as "open-chain-cyclophanes"[62].

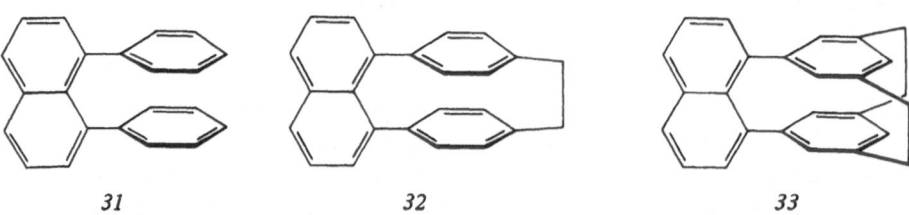

31   32   33

---

[f] The phenyl protons of 1-phenylnaphthalene, for comparison, resonate at δ = 7.38 ppm[61].
[g] By now, the twice-bridged peri-diphenylnaphthalene 33 is under preparation, for reference purposes[63].

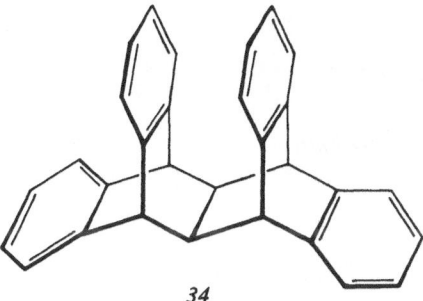

34

2.10. Multiply Bridged Heteraphanes

Triply bridged heteraphane systems, which are in part thia- and aza-analogues of Huberts carbocycles, have been reported. Bond angles and length of trithiaphane *35*[18, 66] as found by X-ray analysis[67], are of normal magnitude and point to only slight molecular strain. The planar benzene rings are separated by 3.2 Å. The site of

35   *36*

one sulfur atom appears to be undefined, indicating rapid inversion of the corresponding CH_2SCH_2-bridge in the crystal (see Fig. 9). Contrary to trisulfide *35*, the bridge protons of which still absorb as a singlet at −110 °C, the more tightly clamped triazaphane *36*[68] shows a temperature dependent ^{1}H−NMR-spectrum: the singlet of the methylene protons splits into an AB-multiplet at −57 °C. The free enthalpy of activation of the corresponding motional process was found $\Delta G_c^* = 13.6$ kcal/mole

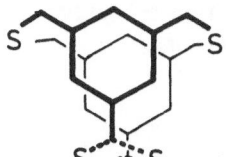

Fig. 9. Disorder of one of the sulfur bridges in crystalline trithiaphane *35*[67]

(56.9 kJ/mole). Flexible long chain thiaphanes *37*[69, 70] and *38*[70], easily synthesized by special dilution techniques, no longer indicate transanular anisotropic effects by their NMR spectra. Thiaphanes *39*[44, 44b, 71] and *40*[44, 44b] synthesized from two molecular building blocks utilizing high dilution techniques can be converted into the hydrocarbons *19* and *21*.

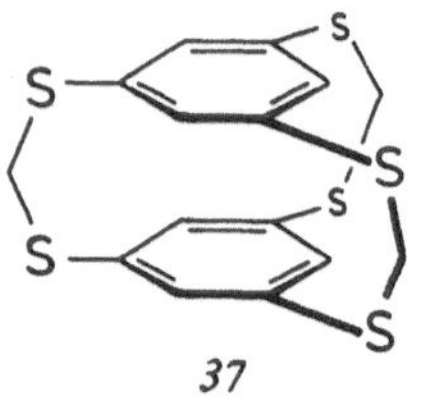

37 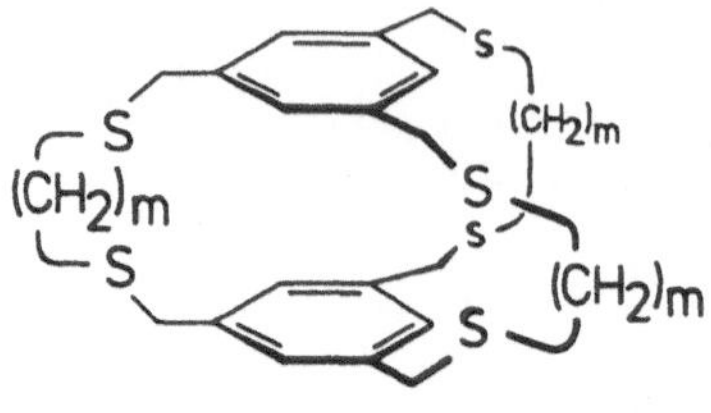   38: m = 1−3

In the ^{1}H−NMR-spectrum of compound 41[44, 44b)] an AA'BB' multiplet is found at the unexpected δ-value of 6.39 ppm; it is assigned to the triphenylethane arene protons, which, according to molecular models, come to lie in the shielding region of an opposing phenylene ring of the triphenylbenzene unit.

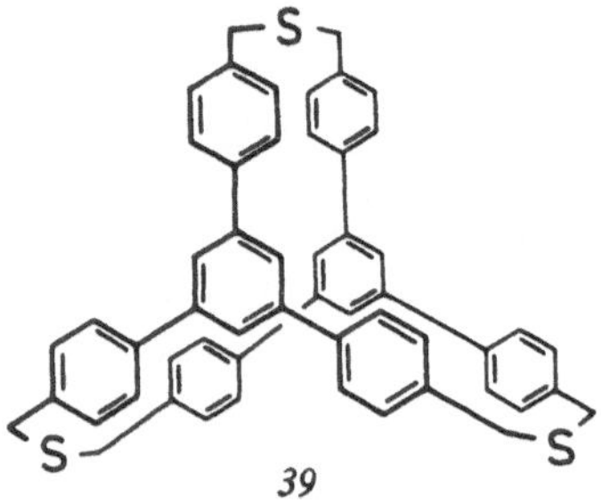

39 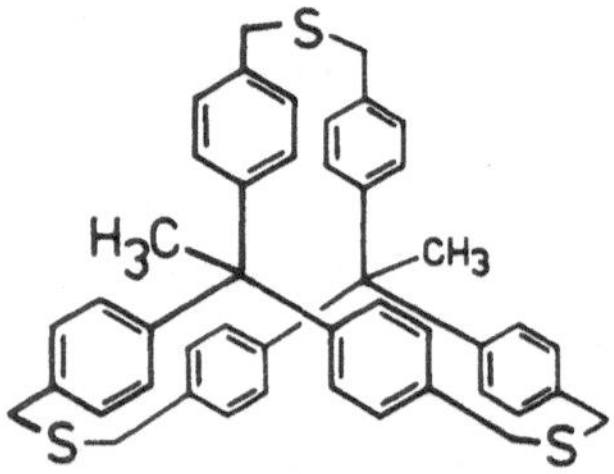   40

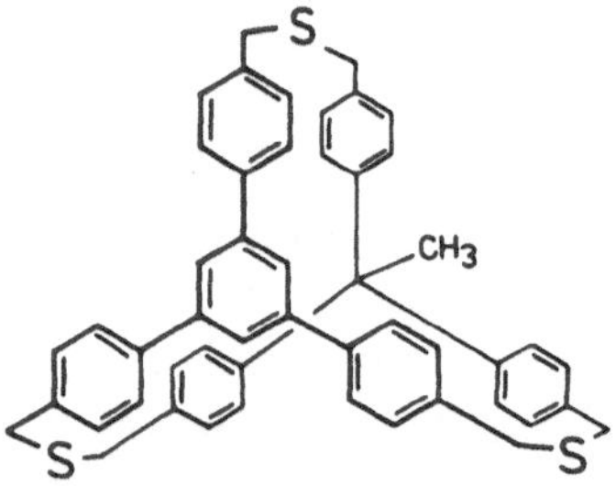

41

Synthesis of [3.3.3]trithiabiphenylophane 42 led to two diastereomers, 42a and 42b[55)]. ^{1}H−NMR-spectra of the biphenyl protons show signals at typical high field strength (δ = 6.2−6.7 ppm), assigned to the asterisked aromatic protons.

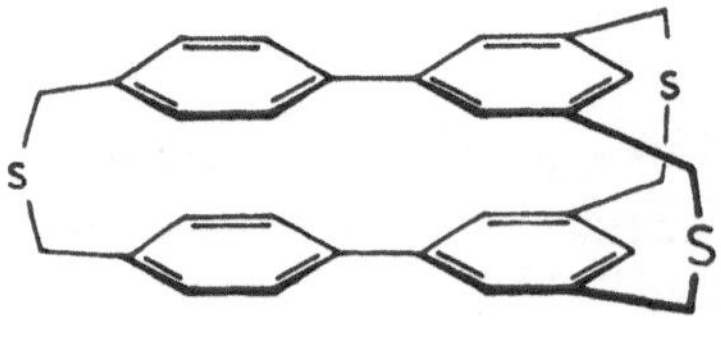

42a 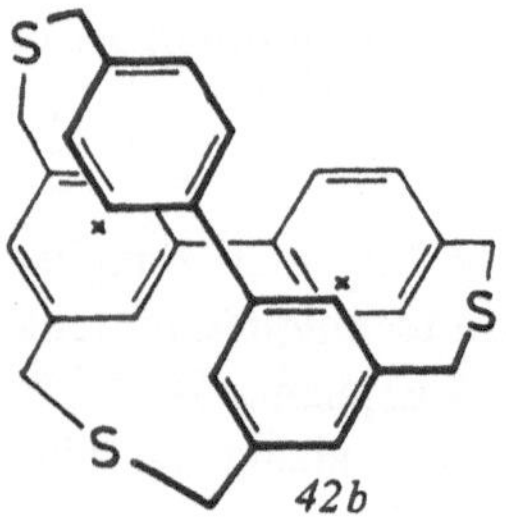   42b

The only products in the syntheses of the quadruply bridged tetrathia[3.3.3.3]-biphenylophanes 43a[71a)] and 43b[71b)] seem to be the cross isomers.

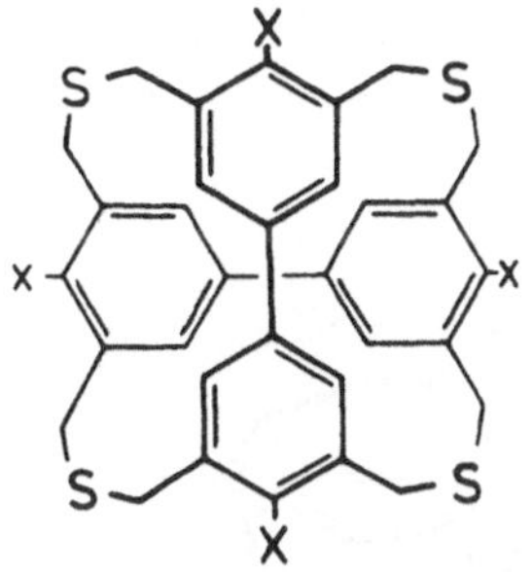

43a: X=H
43b: X=OCH$_3$

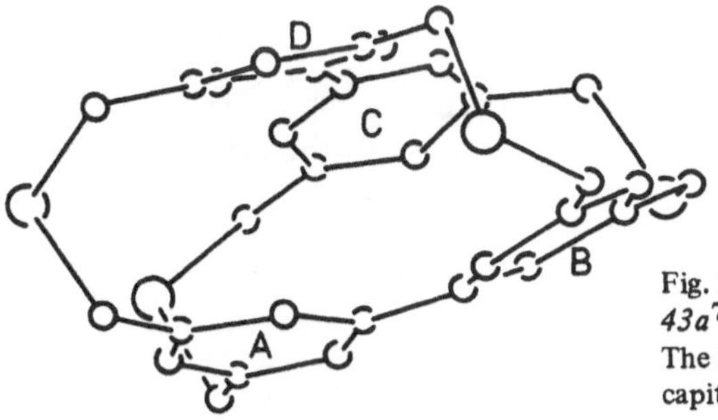

Fig. 10. Geometry of tetrathiaphane 43a[72)].
The benzene rings are marked by capital letters

The structure of 43a was unravelled by X-ray analysis[72)] (Fig. 10). The longer axes of the biphenyl units are bent towards the molecular center by an angle of approximately 20°. In the ^{1}H—NMR-spectrum, the o,o'-proton signals are only slightly shifted upfield despite of their exposure to the shielding zone of the opposite rings. In contrast, quadruple linkage of two tetraphenyl ethene units apparently again yields a mixture of the possible diastereomers 44a and 44b[73)h)].

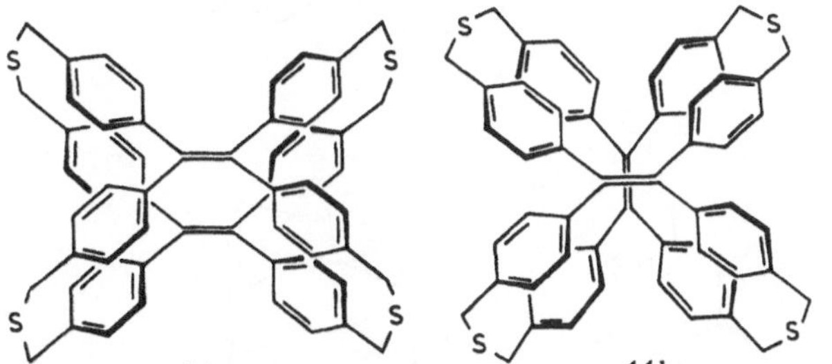

44a   44b

Sulfur extrusion is planned to yield the corresponding carbocyclic [2.2.2.2]-phanes, which are expected from regarding molecular models, to be fixed chiralilly (propeller-like). Hitherto, phane molecules containing more than four hetero bridges

___

h) Isomer ratio 2:3 or 3:2.

are unknown. Attempts to synthesize the hexathiaphane *47* from the benzene derivatives *45* and *46* under various dilution conditions only led to the trithia "monomer"
48[74)i]. Pyrolysis of trisulfone *53*, derived from *48*, might be a possible route to the
still unknown fascinating phane hydrocarbon *55*. In analogy to short-lived *p*- and
o-xylylene yielding [2.2]para-[75)] and orthocyclophane[76)], radialene *54* may dimerize
to *55* under appropriate conditions.

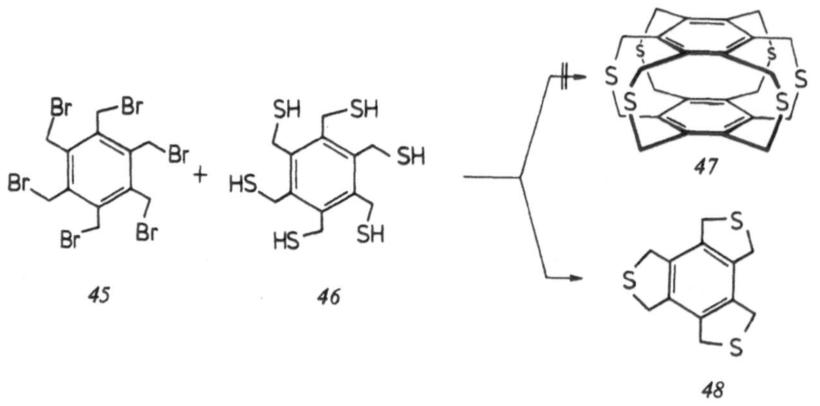

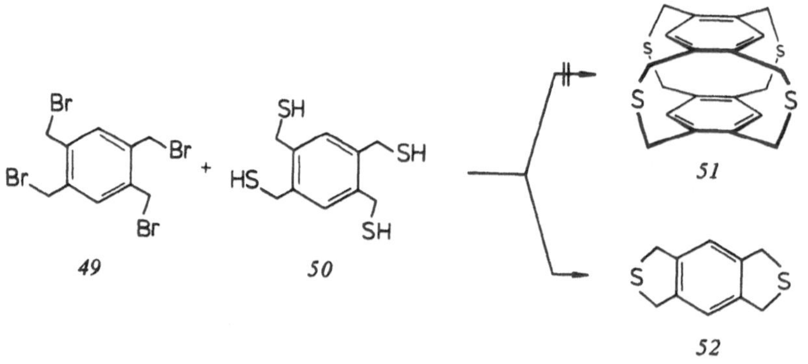

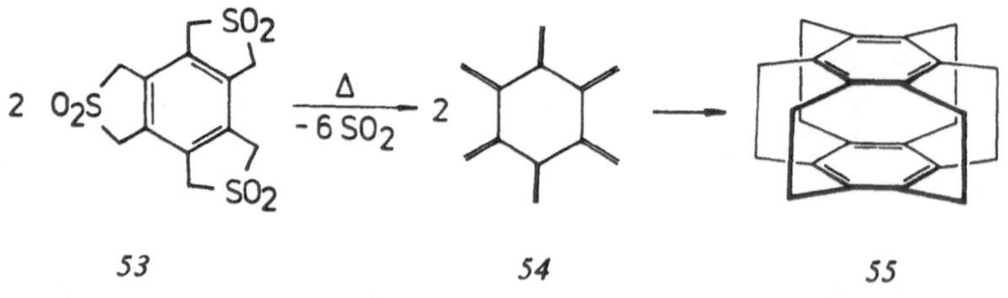

---

i) In analogy, reaction of compounds *49* and *50* did not yield tetrathiaphane *51*, but led to
   52[74)].

3. Multilayered Aromatic Compounds

The first examples of these attractive compounds, the quadruple-layered distereomers *3* and *4*, were synthesized by combining of two [2.2]paracyclophane units *via* appropriate para-xylylene intermediates[16]. *3* is helically chiral (point group D_2; *3a* shows M-, *3b* P-configuration), *4* is meso-configurated (point group C_{2h}). According to X-ray analysis[77], performed later, the two inner benzene rings of the less soluble achiral isomer *4* are fixed in a *twist conformation*, due to the strain imposed alternately on either side (see Fig. 11).

The outer rings are distorted to the expected boat shape. The nearly equidistant arene layers are separated by c. 3.03 Å. Misumi *et al.* reported on the three- to six-fold-layered homologues *56–59*[78a–f, 79)j]. Their well defined rigid geometry allows to estimate the extent of transanular π-π-interactions from UV-spectroscopy. The

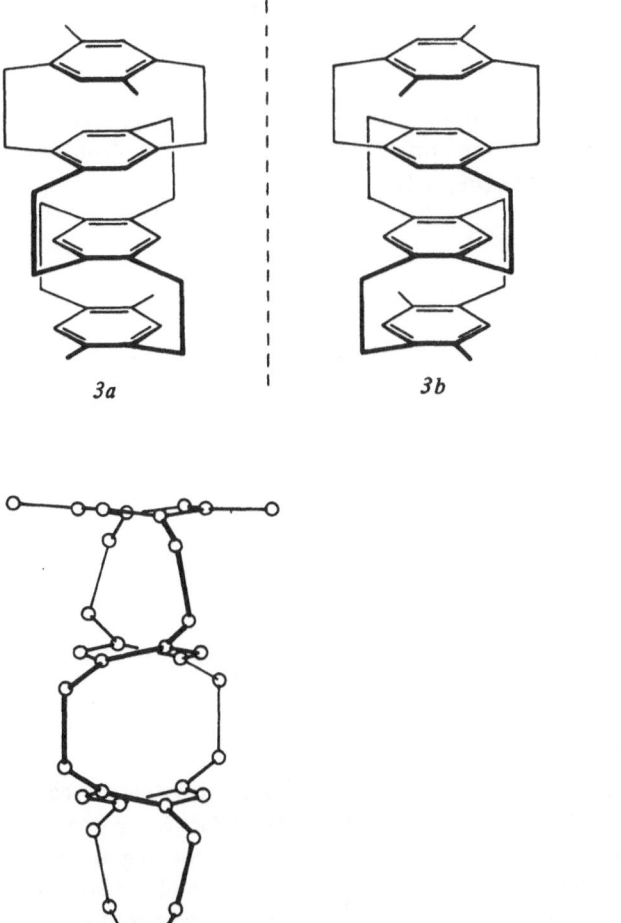

3a   *3b*

Fig. 11. Geometry of phane hydrocarbon *4*[77]

---

j) Also, diastereomeric forms of these phane hydrocarbons have been obtained.

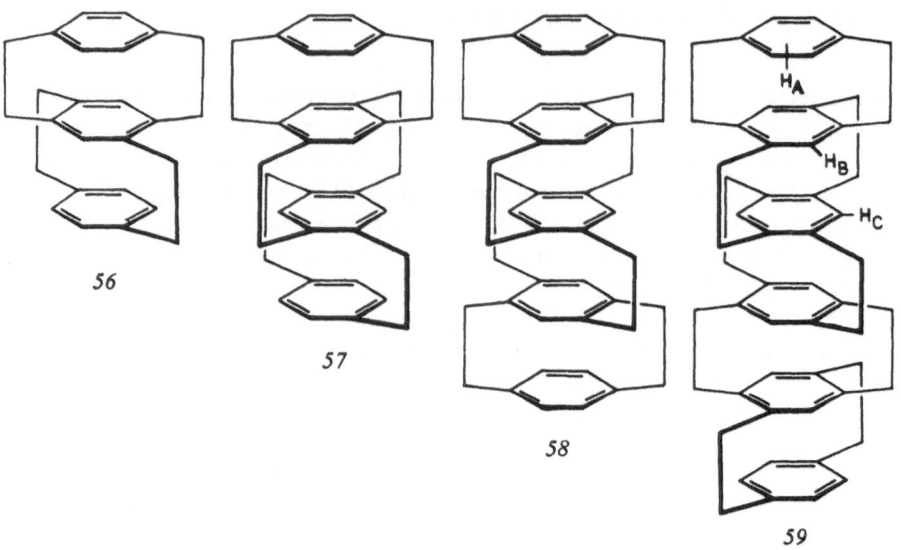

most drastic changes, including a bathochromic shift of the long wavelength maximum by about 45 nm occur on going from double layered [2.2]paracyclophane (*1*) to the triple-layered homologue *56* [80e]) (Fig. 12).

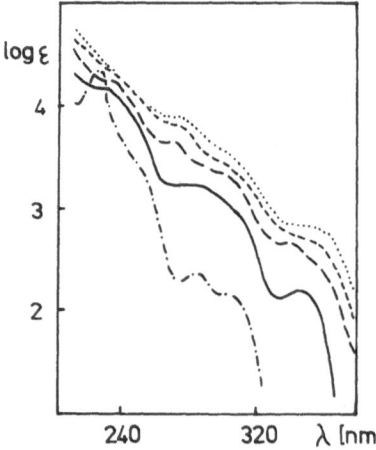

Fig. 12. UV-spectra of compounds *56* (———), *57* (— —), *58* (------), *59* (. . . .), and of [2.2]paracyclophane (*1*) (–.–.–)[79e)]

This striking effect was assigned to extended π-π-interactions in system *56* relative to *1*. However, the absorptional behaviour of the multilayered phanes is apparently also governed by the particular twist deformation of the inner benzene ring(s). This is taught by a comparison between the UV-spectra of [8][8]paracyclophane *60* and of monobridged paracyclophane *61* [80]). Compound *60*, for which a twist conformation of the benzene ring must be assumed, clearly shows transitions at longer wave length than the boat shaped [8]paracyclophane derivative *61*. Hillier and co-workers calculated emission and absorption spectra of *57* on the basis of an extended excita-

20

tion theory also considering intramolecular charge transfer transitions[81]. The calculation at least qualitatively corresponds with the observed bathochromic shifts of *57* relative to [2.2]paracyclophane as a reference.

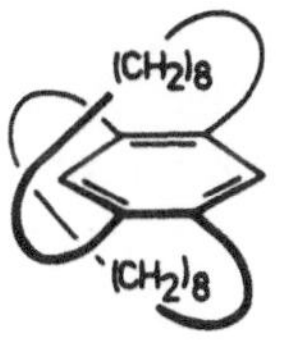

60 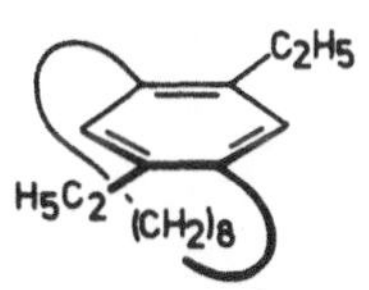   61

Successive adding a new benzene layer in the series *56—59* affects the spectra less and less[78e]. Since these differences ought to scale the transanular electron interactions, and as the absorption spectra of *58* and *59* nearly coincide, the interactions appear to reach their limit in the fivefold-layered phane.

The considerable diamagnetic high field shift of the arene proton signals of the multilayered phanes are rationalized in terms of deshielding increments arising from the next as well as from more distant rings. Hence, maximum high field shift is found for protons of central benzene rings covered on either side by a maximal number of shielding ring currents. *E.g.*, in the sixfold-layered phane *59*, the H_C-protons resonate at $\delta = 4.81$ ppm, and H_B and H_A are progressively shifted to low field strength ($\delta = 5.05$ resp. 6.05 ppm). According to the experimental data, the anisotropic influence of a benzene ring extends significantly at least over the distance of three layers (c. 9 Å)[78f].

Similar to [2.2]paracyclophane[82], *56* rearranges under Friedel-Craft-conditions[83, 84]. Treatment with $AlCl_3$/HCl first results in protonation of one of the *outer* benzene rings. Rearrangement of the thus formed σ-complex and deprotonation yields metaparacyclophane *62*. $SnCl_4$/HCl and other Friedel-Craft acids, however, give rise to the isomers *63* and *64*. Here, apparently, the *inner* benzene ring is subject to protonation and alkyl shift.

In either case, lessening of π-π-repulsion and release of ring strain is supposed to be the driving force of the rearrangement.

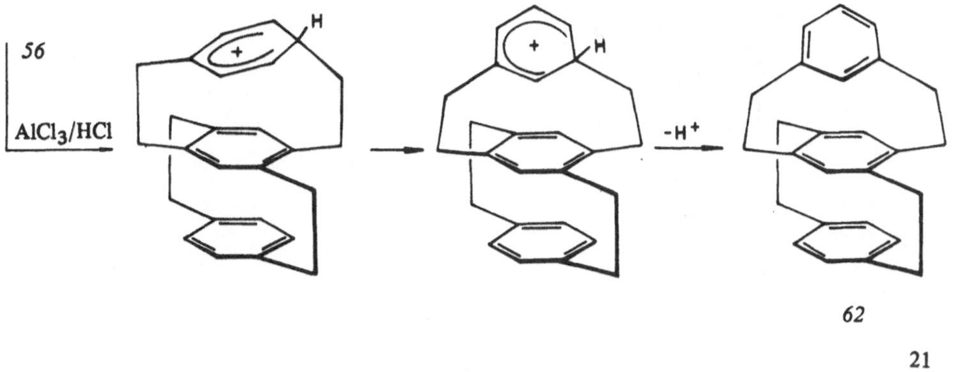

62

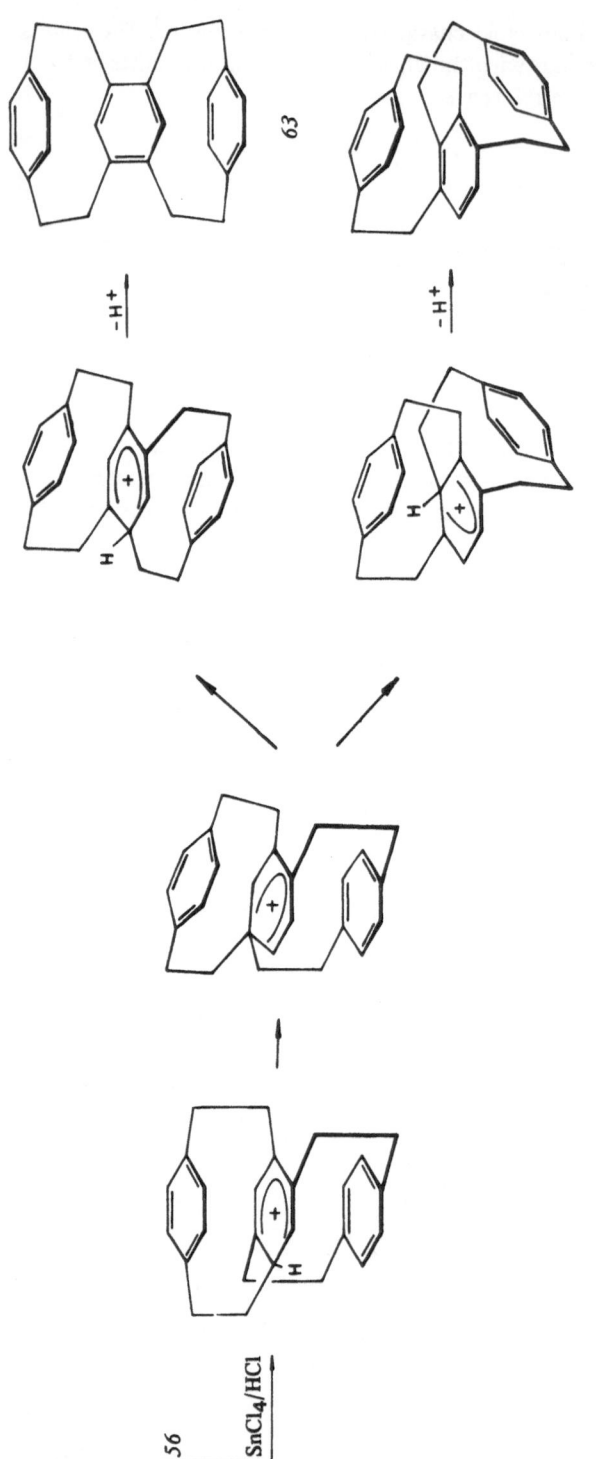

3.1. Multilayered Donor-Acceptor-Complexes

Staab and co-workers[85)] studied intramolecular donor-acceptor complexes, in which the interacting parts are fixed in a face-to-face fashion, *e.g.* the diastereomeric quin-hydrone analogues *65a*[85a)] and *65b*[85b)] with pseudogeminal and pseudoortho orientation of the molecular halves. In the long wave length range (charge transfer bands) the absorption spectra of these compounds are quite distinct. Hence, the dependance of donor-acceptor-interaction on the mutual orientation of the interacting species can be elucidated. In the case of the triple layered[85l)] quinonophanes *66* and *67*[86)] this concept acquires a further dimension. Both diastereomers show charge-transfer-transitions (*66:* λ_{max} = 395 nm, ϵ = 2330; *67:* λ_{max} = 435 nm, ϵ = 320) with strong bathochromic shifts relative to the reference compound *68* (λ_{max} = 340 nm, ϵ = 597)[87)].

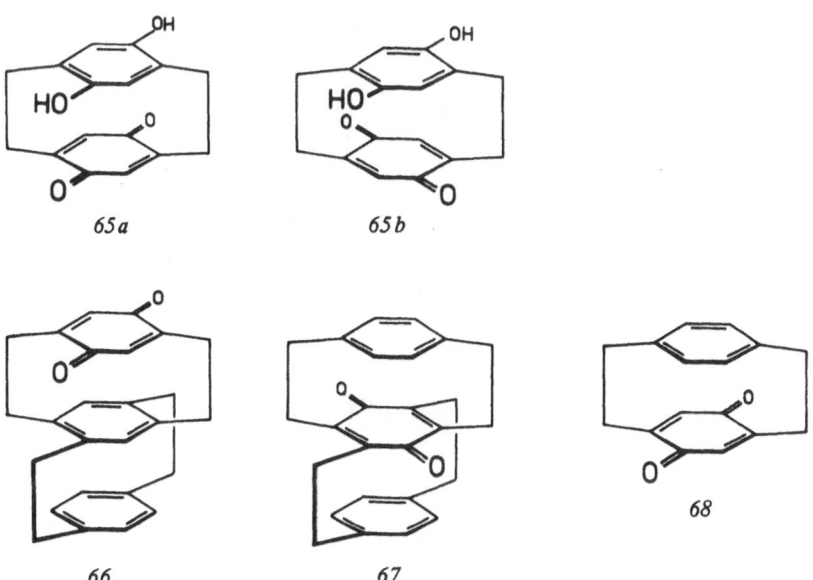

65a   *65b*

66   *67*   *68*

4. Multistepped Aromatic Compounds

[2.2](1,3)(1,3)[2.2](4,6)(1,3)Cyclophane (*5*) forms two diastereomers[17b)]. In the "staircase"-isomer *5a* the benzene rings are arranged "up-up", in the platform isomer *5b* "up-down". Adding more benzene rings increases the number of isomers. In the case of the quadruple-stepped *69*, the three possible forms *69a*, *69b* and *69c* have been isolated. The stability relationships within both isomer groups are remarkable. On heating, *5a* irreversibly forms *5b*, and *69a* as well as *69b* rearrange quantitatively to *69c*[17b)]. Evidently, the maximum number of "up-down" linkages is favored. However, the models predict some destabilization of *5b* and *69c* by repulsion of the

23

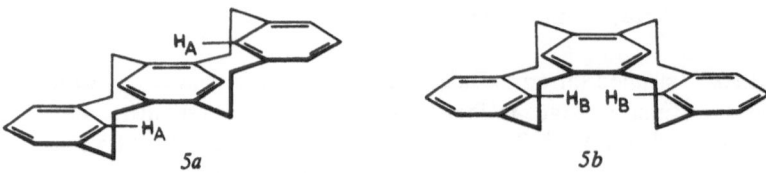

intraanular protons H_B[k]. This inconsistency could be solved as follows: the central benzene ring of diastereomer *5a* shows chair conformation, according to ^{1}H–NMR-studies[88]. On the other hand, the "up-down" geometry of *5b* forces the inner arene ring to adopt a boat-like conformation (cf. Fig. 13).

Fig. 13. Geometries of diastereomers *5a* and *5b*[88]

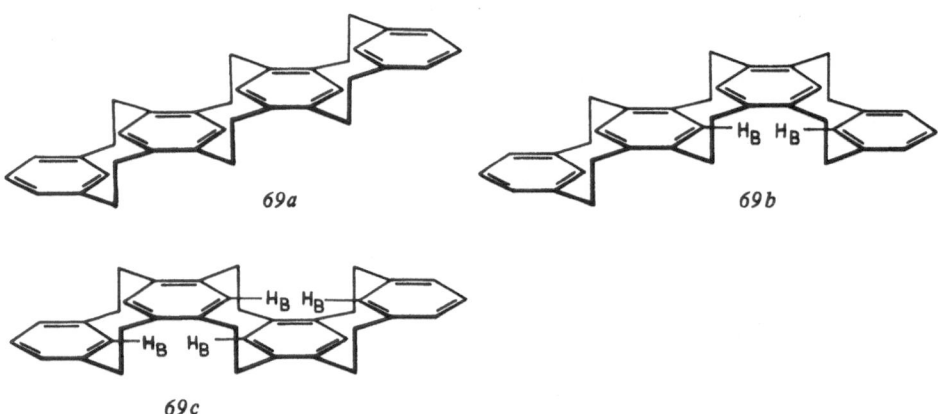

Following SCF–MO-calculations, a boat-like distortion of the benzene ring is obviously favored energetically over chair-geometry, the deformation angles being equal[89]. The different stability is largely due to different values of the resonance integral: in boat-shaped benzene, the atomic *p*-orbitals overlap more effectively and so can gain better mesomeric stabilization[l]. In accordance with experimental evidence, isomer stability is thus predicted to decrease in the order *5b* > *5a* and *69c* > *69b* > *69a*, respectively. Apparently, repulsive interaction between the protons H_B contributes little to the total energy.

---

k) The corresponding ^{1}H–NMR-signals are markedly shifted to lower field with respect to those of the protons H_A[17b] (van der Waals effect[58]).

l) Boat-shaped benzene rings are also found in [2.2]meta-[90] and metaparacyclophanes[91], in helicenes[65] and in a great number of fused arenes (triphenylenes[92], benzo[c]phenanthrene[93]).

Pyridine · HBr$_3$ oxidizes *5a* and *5b* to hydrocarbon *70* via transanular bond-formation[m]; *70* can be dehydrogenated to [2.2]pyrenocyclophane *71*[95, 96]. On treatment of *69a−69c* with the same reagent in each case the inner rings are coupled first[95].

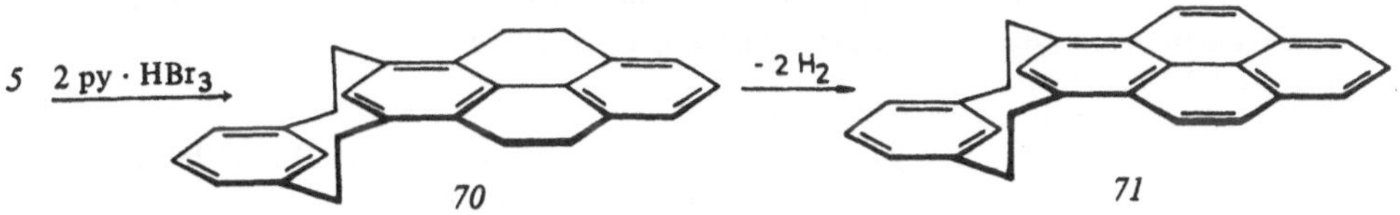

5. Conclusions

The above examples of the unique possibilities of stereochemically fixed aromatic rings in *multibridged, multilayered* and *multistepped* hydrocarbon compounds show that these classes of compounds are most appropriate for the study of intramolecular, steric and electronic interactions, from which too little quantitative relations are known. Again, it follows that further synthetic approaches are not only possible, but also necessary for a more accurate differentiation and analysis of the frequently overlapping electronic and steric effects. More progress in this field is sure to be expected, above all when synthetic and spectroscopic chemists, and physicists adhere to interdisciplinary, collaborative work.

Acknowledgements. We would like to thank the Deutsche Forschungsgemeinschaft and the Fonds der Chemischen Industrie for financial support of our own work; we are grateful to Dipl.-Chem. M. Atzmüller and Miss B. Jendrny for their help in writing up the manuscript, and to P. Koo Tse Mew and W. Offermann for cooperation in translating the paper.

6. References

[1] a) Cram, D. J.: Rec. Chem. Progr. *20*, 71 (1959);
   b) Cram, D. J., Cram, J. M.: Acc. Chem. Res. *4*, 204 (1971)
[2] Vögtle, F., Neumann, P.: Top. Curr. Chem. *48*, 67 (1974); there, a great number of references concerning structure and chemistry of the [2.2]paracyclophane system.
[3] a) Cram, D. J., Steinberg, H.: J. Amer. Chem. Soc. *73*, 5691 (1951);
   b) Cram, D. J., Allinger, N. L., Steinberg, H.: J. Amer. Chem. Soc. *76*, 6132 (1954);
   c) Cram, D. J., Bauer, R. H., Allinger, N. L., Reeves, R. A., Wechter, W. J., Heilbronner E.: J. Amer. Chem. Soc. *81*, 5977 (1959);
   d) Helgeson, R. C., Cram, D. J.: J. Amer. Chem. Soc. *88*, 509 (1966)
[4] Ingraham, L. L.: J. Chem. Phys. *18*, 988 (1950)
[5] McClure, D. S.: Can. J. Chem. *36*, 59 (1958)

[m] Compare the analogous behaviour of [2.2]metacyclophane with respect to electrophilic reagents[94].

6) a) Koutecký, J., Paldus, J.: Collection Czech. Chem. Commun. *27*, 599 (1962);
   b) Koutecký, J., Paldus, J.: Tetrahedron *19*, Suppl. 2, 201 (1963);
   c) Paldus, J.: Collection Czech. Chem. Commun. *28*, 1110 (1963)
7) a) Ron, A., Schnepp, O.: J. Chem. Phys. *37*, 2540 (1962); *44*, 19 (1966);
   b) Vala, Jr., M. T., Haebig, J., Rice, S. A.: J. Chem. Phys. *43*, 886 (1965);
   c) Vala, Jr., M. T., Hillier, I. H., Rice, S. A., Jortner, J.: J. Chem. Phys. *44*, 23 (1966)
8) Longworth, J. W., Bovey, F. A.: Biopolymers *4*, 1115 (1966)
9) Gleiter, R.: Tetrahedron Lett. *1969*, 4453
10) El Ayed, M. A.: Nature *197*, 481 (1963)
11) a) Cram, D. J., Wechter, W. J., Kierstead, R. W.: J. Am. Chem. Soc. *80*, 3126 (1958);
    b) Reich, H. J., Cram, D. J.: J. Am. Chem. Soc. *90*, 1365 (1968);
    c) Reich, H. J., Cram, D. J.: J. Am. Chem. Soc. *91*, 3505 (1969);
    d) Reich, H. J., Cram, D. J.: J. Am. Chem. Soc. *91*, 3527 (1969)
    e) For transanular interactions in 4-verdazyl[2.2]paracyclophanes see: Neugebauer, F. A., Fischer, H.: Tetrahedron Lett. *1977*, 3345
12) Hope, H., Bernstein, J., Trueblood, K. N.: Acta Crystallogr. *B28*, 1733 (1972)
13) Wynberg, H., Nieuwpoort, W. C., Jonkman, H. T.: Tetrahedron Lett. *1973*, 4623; further references there
14) Binsch, G.: Naturwissenschaften *60*, 369 (1973)
15) Otsubo, T., Mizogami, S., Otsubo, I., Tozuka, Z., Sakagami, A., Sakata, Y., Misumi, S.: Bull. Chem. Soc. Jap. *46*, 3519 (1973)
16) a) Longone, D. T., Chow, H. S.: J. Am. Chem. Soc. *86*, 3898 (1964);
    b) Longone, D. T., Chow, H. S.: J. Am. Chem. Soc. *92*, 994 (1970)
17) a) Umemeto, T., Otsubo, T., Sakata, Y., Misumi, S.: Tetrahedron Lett. *1973*, 593;
    b) Umemeto, T., Otsubo, T., Misumi, S.: Tetrahedron Lett. *1974*, 1573
18) Boekelheide, V., Hollins, R. A.: J. Am. Chem. Soc. *92*, 3512 (1970); in May 1969 a program has been started by Vögtle, F., for the synthesis of *2* and its mono-, di- and triene with support of the Deutsche Forschungsgemeinschaft
19) Hanson, A. W., Röhrl, M.: Acta Crystallogr. *B28*, 2287 (1972)
20) Gantzel, P. K., Trueblood, K. N.: Acta Crystallogr. *18*, 958 (1965)
21) Coulter, C. L., Trueblood, K. N.: Acta Crystallogr. *16*, 667 (1963)
22) Robertson, J. M.: Organic crystals and molecules. Ithaca-New York: Cornell University Press 1953
23) Boekelheide, V., Hollins, R. A.: J. Am. Chem. Soc. *95*, 3201 (1973)
24) Hanson, A. W., Röhrl, M.: Acta Crystallogr. *B28*, 2032 (1972)
25) Hanson, A. W.: Acta Crystallogr. *15*, 956 (1962)
26) Hanson, A. W., Huml, K.: Acta Crystallogr. *B27*, 459 (1971)
27) See f.e.: a) Cram, D. J., Dalton, C. K., Knox, G. R.: J. Am. Chem. Soc. *85*, 1088 (1963);
    b) Cram, D. J., Helgeson, R. C.: J. Am. Chem. Soc. *88*, 3515 (1966);
    c) Wasserman, H. H., Keehn, P. M.: J. Am. Chem. Soc. *91*, 2374 (1969);
    d) Toyoda, T., Otsubo, I., Sakata, Y., Misumi, S.: Tetrahedron Lett. *1972*, 1731;
    e) Staab, H. A., Haenel, M.: Chem. Ber. *106*, 2190 (1973)
28) Newton, M. G., Walter, T. J., Allinger, N. L.: J. Am. Chem. Soc. *95*, 5652 (1973)
29) Wolf, A. D., Kane, V. V., Levin, R. H., Jones, Jr., M.: J. Am. Chem. Soc. *95*, 1680 (1973)
30) Allinger, N. L., Walter, T. J., Newton, M. G.: J. Am. Chem. Soc. *96*, 4588 (1974)
31) Kane, V. V., Wolf, A. D., Jones, Jr., M.: J. Am. Chem. Soc. *96*, 2634 (1974)
32) Allinger, N. L., Sprague, J. T., Liljefors, T.: J. Am. Chem. Soc. *96*, 5100 (1974)
33) As to the photoelectron spectrum of triene *6;* see: Boschi, R., Schmidt, W.: Angew. Chem. *85*, 408 (1973); Angew. Chem., Int. Ed. Engl. *12*, 402 (1973)
34) Truesdale, E. A., Cram, D. J.: J. Am. Chem. Soc. *95*, 5825 (1973); see also: Trampe, S., Menke, K., Hopf, H.: Chem. Ber. *110*, 371 (1977)
35) Nakazaki, M., Yamamoto, K., Miura, Y.: J. Chem. Soc. Chem. Commun. *1977*, 206
36) Hubert, A. J., Dale, J.: J. Chem. Soc. *1965*, 3160
37) a) Hubert, A. J.: J. Chem. Soc. C, *1967*, 6; see also:
    b) Hubert, A. J., Hubert, M.: Tetrahedron Lett. *1966*, 5779;
    c) Hubert, A. J.: J. Chem. Soc. C, *1967*, 11; J. Chem. Soc. C, *1967*, 13

[38] a) Cram, D. J., Bauer, R. H.: J. Am. Chem. Soc. *81*, 5971 (1959);
b) Singer, L. A., Cram, D. J.: J. Am. Chem. Soc. *85*, 1080 (1963);
c) Sheehan, M., Cram, D. J.: J. Am. Chem. Soc. *91*, 3553 (1969)

[39] Langer, E., Lehner, H.: Tetrahedron *29*, 375 (1973)

[40] Dewar, M. J. S., Thompson, Jr., C. C.: Tetrahedron Suppl. *7*, 97 (1966)

[41] Gray, R., Boekelheide, V.: Angew. Chem. *87*, 138 (1975); Angew. Chem., Int. Ed. Engl. *14*, 107 (1975)

[42] Gilb, W., Menke K., Hopf, H.: Angew. Chem. *89*, 177 (1977); Angew. Chem., Int. Ed. Engl. *16*, 191 (1977)

[43] a) Stephens, R. D.: J. Org. Chem. *38*, 2260 (1973);
b) Gladysz, J. A., Fulcher, J. G., Lee, S. J., Bocarsley, A. B.: Tetrahedron Lett. *1977*, 3421

[44] a) Vögtle, F., Hohner, G.: Angew. Chem. *87*, 522 (1975); Angew. Chem., Int. Ed. Engl. *14*, 497 (1975);
b) Hohner, G., Vögtle, F.: Chem. Ber. *110*, 3052 (1977)

[45] Mislow, K., Gust, D., Finocchiaro, P., Boettcher, R. J.: Topics Curr. Chem. *47*, 1 (1974)

[46] Farina, M.: Tetrahedron Lett. *1963*, 2097; Allegra, G., Farina, M. Immirzi, A., Colombo, A., Rossi, U., Broggi, R., Natta, G.: J. Chem. Soc. B *1967*, 1020; Farina, M., Audisio, G.: Tetrahedron *26*, 1827 (1970); Farina, M., Audisio, G.: Tetrahedron *26*, 1839 (1970)

[47] Hill, R. K., Ladner, D. W.: Tetrahedron Lett. *1975*, 989

[48] Morandi, C., Mantica, E., Botta, D., Gramegna, M. T., Farina, M.: Tetrahedron Lett. *1973*, 1141

[49] See also: Wittig, G., Schoch, W.: Liebigs Ann. Chem. *749*, 38 (1971)

[50] Vögtle, F., Winkel, J.: in preparation

[51] Lankamp, H., Nauta, W. Th., MacLean, C.: Tetrahedron Lett. *1968*, 249

[52] Staab, H. A., Brettschneider, H., Brunner, H.: Chem. Ber. *103*, 1101 (1970)

[53] See: McBridge, J. M.: Tetrahedron *30*, 2009 (1974)

[54] Gerson, F., Martin, Jr., W. B.: J. Am. Chem. Soc. *91*, 1883 (1969)

[55] Vögtle, F., Steinhagen, G.: Chem. Ber. *111*, 205 (1978); Grütze, J., Vögtle, F.: Chem. Ber. *110*, 1978 (1977); there further references

[56] See Vögtle, F.: Angew. Chem. *81*, 258 (1969); Angew. Chem., Int. Ed. Engl. *8*, 274 (1969)

[57] Vögtle, F., Offermann, W.: unpublished results of continuing work

[58] Pople, J. A., Schneider, W. G., Bernstein, H. J.: High-resolution nuclear magnetic resonance. New York-Toronto-London: McGraw-Hill Book Co., Inc. 1959, p. 253

[59] Högberg, H. E., Thulin, B., Wennerström, O.: Tetrahedron Lett. *1977*, 931; compare loc. cit.[37c]

[60] Ogilvie, R. A.: Ph. D. Thesis, Massachusetts Institute of Technology, 1971

[61] House, H. O., Magin, R. W., Thompson, H. W.: J. Org. Chem. *28*, 2403 (1963)

[62] Bieber, W., Vögtle, F.: Angew. Chem. *89*, 199 (1977); Angew. Chem., Int. Ed. Engl. *16*, 175 (1977)

[63] Vögtle, F., Wingen, R.: unpublished results

[64] Cristol, S. J., Lewis, D. C.: J. Am. Chem. Soc. *89*, 1476 (1967)

[65] Cf. Martin, R. H.: Angew. Chem. *86*, 727 (1974); Angew. Chem., Int. Ed. Engl. *13*, 649 (1974)

[66] Vögtle, F.: Liebigs Ann. Chem. *735*, 193 (1970)

[67] Hanson, A. W., Macaulay, E. W.: Acta Crystallogr. *B28*, 1255 (1972)

[68] Vögtle, F., Neumann, P.: J. Chem. Soc. Chem. Commun. *1970*, 1464

[69] Vögtle, F., Lichtenthaler, R. G.: Tetrahedron Lett. *1972*, 1905

[70] Lichtenthaler, R. G., Vögtle, F.: Chem. Ber. *106*, 1319 (1973)

[71] a) Vögtle, F., Hohner, G., Weber, E.: J. Chem. Soc. Chem. Commun. *1973*, 366;
b) Vögtle, F., Heidan, C.: unpublished results; Heidan, C.: Zulassungsarbeit, Universität Würzburg 1975

[72] Matsumoto, K., Nowacki, W.: Z. Krist. *141*, 260 (1975)

[73] Wester, N.: Diplomarbeit, Universität Bonn, 1977; Vögtle, F., Wester, N.: Liebigs Ann. Chem. *1978*, in press

[74] Neumann, P.: Dissertation, Universität Heidelberg, 1973

[75] Cava, M. P., Deana, A. A.: J. Am. Chem. Soc. *81*, 4266 (1959)

[76] Winberg, H. E., Fawcett, F. S., Mochel, W. E., Theobald, C. W.: J. Am. Chem. Soc. *82*, 1428 (1960)

[77] Mizuno, H., Nishiguchi, K., Otsubo, T., Misumi, S., Morimoto, N.: Tetrahedron Lett. *1972*, 4981

[78] a) Otsubo, T., Mizogami, S., Sakata, Y., Misumi, S.: J. Chem. Soc. Chem. Commun. *1971*, 678;
   b) Otsubo, T., Mizogami, S., Sakata, Y., Misumi, S.: Tetrahedron Lett. *1971*, 4803;
   c) Otsubo, T., Tozuka, Z., Mizogami, S., Sakata, Y., Misumi, S.: Tetrahedron Lett. *1972*, 2972;
   d) Otsubo, T., Mizogami, S., Sakata, Y., Misumi, S.: Tetrahedron Lett. *1973*, 2457;
   e) Otsubo, T., Mizogami, S., Otsubo, I., Tozuka, Z., Sakagami, A., Sakata, Y., Misumi, S.:
      Bull. Chem. Soc. Jap. *46*, 3519 (1973); for strain energy calculations see: Nishiyama, K.,
      Sakiyama, M., Seki, S.: Tetrahedron Lett. *1977*, 3739
   f) Otsubo, T., Mizogami, S., Sakata, Y., Misumi, S.: Bull. Chem. Soc. Jap. *46*, 3831 (1973);
      see also: Mizogami, S., Otsubo, T., Sakata, Y., Misumi, S.: Tetrahedron Lett. *1971*, 2791;
      Toyoda, T., Iwama, A., Sakata, Y., Misumi, S.: Tetrahedron Lett. *1975*, 3203

[79] See also: Nakazaki, M., Yamamoto, K., Tanaka, S., Kanetani, H.: J. Org. Chem. *42*, 287 (1977)

[80] Nakazaki, M., Yamamoto, K., Tanaka, S.: Tetrahedron Lett. *1971*, 341

[81] Hillier, I. H., Glass, L., Rie, S. A.: J. Am. Chem. Soc. *88*, 5063 (1966)

[82] Hefelfinger, D. T., Cram, D. J.: J. Am. Chem. Soc. *93*, 4754 (1971)

[83] Horita, H., Kannen, N., Otsubo, T., Misumi, S.: Tetrahedron Lett. *1974*, 501

[84] Compare also: Kannen, N., Otsubo, T., Sakata, Y., Misumi, S.: Bull. Chem. Soc. Jap. *49*, 3203 (1976); Bull. Chem. Soc. Jap. *49*, 3207 (1976)

[85] a) Rebafka, W., Staab, H. A.: Angew. Chem. *85*, 831 (1973); Angew. Chem., Int. Ed. Engl. *12*, 776 (1973);
   b) Rebafka, W., Staab, H. A.: Angew. Chem. *86*, 234 (1974); Angew. Chem., Int. Ed. Engl. *13*, 203 (1974);
   c) Staab, H. A., Herz, C. P., Henke, H.-E.: Tetrahedron Lett. *1974*, 4393;
   d) Staab, H. A., Haffner, H.: Tetrahedron Lett. *1974*, 4397;
   e) Vogler, H., Ege, G., Staab, H. A.: Tetrahedron *31*, 2441 (1975);
   f) Staab, H. A., Herz, C. P.: Angew. Chem. *89*, 406 (1977); Angew. Chem., Int. Ed. Engl. *16*, 392 (1977);
   g) Herz, C. P., Staab, H. A.: Angew. Chem. *89*, 407 (1977); Angew. Chem., Int. Ed. Engl. *16*, 394 (1977);
   h) Staab, H. A., Rebafka, W.: Chem. Ber. *110*, 3333 (1977);
   i) Staab, H. A., Herz, C. P., Henke, H.-E.: Chem. Ber. *110*, 3351 (1977);
   j) Staab, H. A., Haffner, H.: Chem. Ber. *110*, 3359 (1977);
   k) Staab, H. A., Taglieber, V.: Chem. Ber. *110*, 3367 (1977);
   l) see also Staab, H. A., *et al.*: in press

[86] a) Tatemitsu, H., Otsubo, T., Sakata, Y., Misumi, S.: Tetrahedron Lett. *1975*, 3059; see also:
   b) Yoshida, M., Tatemitsu, H., Sakata, Y., Misumi, S.: J. Chem. Soc. Chem. Commun. *1976*, 587;
   c) Horita, H., Otsubo, T., Sakata, Y., Misumi, S.: Tetrahedron Lett. *1976*, 3899;
   d) Transanular π-electron interactions in "Multilayered Thiophenophanes and Furanophanes":
      Otsubo, T., Mizogami, S., Osaka, N., Sakata, Y., Misumi, S.: Bull. Chem. Soc. Jap. *50*, 1841, 1858 (1977)

[87] Cram, D. J., Day, A. C.: J. Org. Chem. *31*, 1227 (1966)

[88] Lehner, H.: Monatshefte Chem. *107*, 565 (1976)

[89] Iwamura, H., Kihara, H., Misumi, S., Sakata, Y., Umemeto, T.: Tetrahedron Lett. *1976*, 615.

[90] Cf.: Vögtle, F., Neumann, P.: Angew. Chem. *84*, 75 (1972); Angew. Chem., Int. Ed. Engl. *11*, 73 (1872)

[91] Cf.: Vögtle, F., Neumann, P.: Chimia *26*, 64 (1972)

[92] Ferguson, G. in: Advances in physical organic chemistry. Gold, V. (ed.). London: Academic Press 1963, Vol. 1, p. 203

[93] Reid, C.: J. Mol. Spec. *1*, 18 (1957)

94) Sato, T., Wakabayashi, M., Okamura, A., Amada, T., Hata, K.: Bull. Chem. Soc. Jap. *40*, 2363 (1967)
95) Umemeto, T., Kawashima, T., Sakata, Y., Misumi, S.: Tetrahedron Lett. *1975*, 463; Chem. Lett. *1975*, 837
96) Umemeto, T., Kawashima, T., Sakata, Y., Misumi, S.: Tetrahedron Lett. *1975*, 1005

Received November 15, 1977

Isotope Effects in Hydrogen Atom Transfer Reactions

Edward S. Lewis

Department of Chemistry, Rice University, P. O. Box 1892, Texas 77001, U. S. A.

Table of Contents

Introduction

Isotope effects are of interest because their existence tells us about features of mechanism, their magnitude is an important test of the most fundamental of theories of kinetics, and they also bear on the more empirical understanding of reaction rates. Hydrogen atom transfers are one of the most general reactions of free radicals in organic systems. They are about the least complicated substitution reaction for the theoretician, so that more progress has been made in understanding them than for any other chemical reaction, even though complete solutions are not yet in hand. The hydrogen atom transfer shares with the ionic proton and hydride transfers the distinction of being the reactions with the largest isotope effects (because of the low mass of hydrogen), and the large magnitude allows significant experimental measurements to be made without the extreme precision required for other isotope effect measurements.

Unlike the proton or hydride transfers, the atom transfers need not involve ionic species, although ionic species can also undergo this reaction. Thus we are not of necessity concerned with the enormous solvation energies, the various levels of solvation, and the enormous quantitative differences between gas phase and solution chemistry that add complexity to the extensive studies of proton transfer reactions[1] and the much less studied hydride transfer reactions. Another interesting contrast is that hydrogen atom transfers, even of very low activation energy, appear to be slow enough so that problems of diffusion control do not arise.

In this article, emphasis will be on hydrogen atom transfers of the general form of reaction (1) in which either A· or X· or both are organic

$$A· + H:X \longrightarrow A:H + ·X \tag{1}$$

free radicals. No further mention will be made of cases in which either A· or HX are ionic, nor of cases in which HX or HA are themselves free radicals. Emphasis will be on reactions in solution, where there is no problem of long-lived excited vibrational states, although most of the few reactions studied in both phases go at comparable rates and with comparable isotope effects in solution and in the gas phase[2].

Theoretical Basis of Hydrogen Isotope Effects

The majority of work on isotope effects in large molecules has been based on absolute reaction rate theory. Treatments of great generality applicable to effect of any kind of isotopic substitution have been developed[3]. In dealing with primary hydrogen isotope effects, that is those in which the isotopic change is in the hydrogen atom being transferred, considerable simplifications can be introduced because the hydrogen nucleus is so much lighter than the other nuclei normally present. A further great simplification is the limitation of the problem to a one-dimensional three particle model. This model appears to be applicable to both proton transfers and hydrogen atom transfers for which linear transition states are favored[4], but may be grossly in error for hydride transfer reactions, where the transition states may well be triangular[5]. Curiously, the one-dimensional treatment succeeds best when the reagent HX is three-dimensional, and is much less effective when HX is two dimensional triatomic (for example H_2S) or one-dimensional diatomic (for example HBr)[6]. This apparent anomaly is a consequence of the fact that the success of the one-dimensional treatment is based on the cancellation of two bending modes of vibration in the reagent HX and the transition state A—H—X, but H_2S has only one such mode, and HBr has none.

The linear three particle system A—H—X confined to one dimension has in general two fundamental vibrations, both stretching modes, analogous to the two stretching modes of carbon dioxide:

$$\overset{\leftarrow}{O}=C=\vec{O} \text{ and } \overset{\leftarrow}{O}=\overset{\updownarrow}{C}=\overset{\leftarrow}{O}$$

However, because A—H—X is a transition state, only one of these modes has a positive restoring force, its fundamental frequency is sometimes called $\nu_1^{\ddagger}$, the other corresponds to the motion along the reaction coordinate, this motion results in a decrease in energy and can be described as having a negative restoring force constant, and an imaginary frequency $i\nu_3^{\ddagger}$.

The calculation of the isotope effect in terms of this simplified model is then given by the Eq. (2). In this ν_{HX} is the one (stretching) frequency of HX, and $\nu_{1H}^{\ddagger}$ is the real stretching frequency of the transition

$$k_H/k_D = (Q_H/Q_D) e^{\frac{h}{2\,RT} [(\nu_{HX} - \nu_{DX}) - (\nu_{1H}^{\ddagger} - \nu_{1D}^{\ddagger})]} \tag{2}$$

state A—H—X, and $\nu_{1D}^{\ddagger}$ is the corresponding frequency of A—D—X, and Q_H and Q_D are tunnel corrections on k_H and k_D respectively. Since ν_H is always greater than ν_D, the isotope effect is largest when the second parenthetical term disappears, either because $\nu_{1H}^{\ddagger}$ becomes nearly equal to $\nu_{1D}^{\ddagger}$, or because both become so small that the difference is negligible. It is convenient to define the "maximum" isotope effect, in Eq. (3), which

$$(k_H/k_D)_{max} = \exp\left[\frac{h}{2\,RT} (\nu_{HX} - \nu_{DX})\right] \tag{3}$$

(except for the tunnel correction) is so achieved. It is convenient to express this as

$$(k_H/k_D)_{max} = \exp(0.2107\,\nu_{HX}/T),$$

in which ν_{HX} is in cm^{-1}, this carries the assumption that X is much heavier than H, as well as that the vibration is harmonic. The harmonic oscillator approximation is also included in Eq. 2, as is the assumption that the temperature is too low for any significant population of excited states.

In Eq. (2), variation in isotope effect for varying A· can be attributed only to variation in Q_H/Q_D or to variation in $\nu_{1H}^{\ddagger} - \nu_{1D}^{\ddagger}$, and we shall consider these in

33

reverse order. Westheimer[7] first pointed out that for a rather special potential surface for which $\nu_3^{\ddagger} = 0$, a transition state with equal force constants between H and A, and between H and X would have $\nu_{1H}^{\ddagger} = \nu_{1D}^{\ddagger}$, and hence the maximum isotope effect would be achieved. The basis of this argument is that the hydrogen nucleus does not move in this vibration, so that the frequency is independent of its mass. However, with an extremely unsymmetrical reagent-like[8] transition state this is no longer the case; a better treatment is that $\nu_{1H}^{\ddagger} \cong \nu_{HX}$, $\nu_{1D}^{\ddagger} \cong \nu_{DX}$, and (since as will be illustrated below, Q_H/Q_D will be unity) $k_H/k_D \cong 1$.

This variation in the isotope effect, due to variation in the isotope sensitivity of $\nu_1^{\ddagger}$, has been called the Westheimer symmetry effect, and it will be one of the central ideas of this paper. However, in connection with proton transfers it has been attacked, for when more realistic potential energy surfaces are used (that is $\nu_3^{\ddagger} > 0$), a much greater degree of force constant asymmetry is required to get a much reduced isotope effect[9–11]. Bell has therefore suggested that much of the observed variation in the isotope effect is due to variation in the tunnel correction, Q_H/Q_D.

The tunnel correction is an artificial device to correct an artificial treatment, in which the real vibrations of the transition state are treated quantum mechanically, but the passage along the reaction coordinate over the barrier is treated classically. This correction has the property that $Q_H > Q_D > 1$; it thus can only increase the isotope effect. It is dependent on several factors[12] including especially $\nu_3^{\ddagger}$, and the Q_H/Q_D can only be much greater than unity if $\nu_{3H}^{\ddagger} > \nu_{3D}^{\ddagger}$. It can be shown[11,13] that if A and X are much heavier than hydrogen, that

$$\nu_{1H}^{\ddagger} \nu_{3H}^{\ddagger} / \nu_{1D}^{\ddagger} \nu_{3D}^{\ddagger} = 2^{1/2} \qquad (4)$$

Thus if the isotope effect is small, because $\nu_{1H}^{\ddagger} - \nu_{1D}^{\ddagger}$ is nearly as large as $\nu_{HX} - \nu_{DX}$, in Eq. (2), then $\nu_{1H}^{\ddagger}$ is nearly $2^{1/2} \nu_{1D}^{\ddagger}$, and hence $\nu_{3H}^{\ddagger} \approx \nu_{3D}^{\ddagger}$ and the tunnel correction is small[14]. However, the tunnel correction is largest when $\nu_{1H}^{\ddagger} = \nu_{1D}^{\ddagger}$, when the uncorrected isotope effect approaches the maximum value. In the neighborhood of room temperature, tunnel correction as large as $Q_H/Q_D = 5$ have been suspected and values around 2 are not very unusual[12]. The estimation of Q_H/Q_D from experimental data is uncertain, and values of $Q_H/Q_D < 1.5$ probably escape detection. The conclusion from Eq. (4) that small isotope effects well below the maximum will have small tunnel corrections is firm for the one-dimensional model and is not likely to be wrong for real three-dimensional systems, as long as the transition state is linear. However, tunnel corrections may be a large component of isotope effects in hydride transfers with triangular transition states[15].

An alternative method of estimating isotope effects is interesting in that it also predicts a maximum isotope effect for a symmetrical situation and smaller isotope effects for the unsymmetrical cases. Marcus[16] has presented a quadratic equation for proton (and electron transfers) relating the free energy of activation ($\Delta G^{\ddagger}$) to an energy term for getting the reagents in position to react (w^r), the free energy change after this proximity is achieved accompanying the proton transfer process ($\Delta G^{\circ\prime}$), and the so-called intrinsic barrier ($\lambda/4$), which is the barrier when $\Delta G^{\circ\prime} = 0$. Marcus has pointed out the value of this in connection with isotope effects, which has been exploited, especially by Kreevoy and Oh[17], to explain variation of isotope effects in proton transfers. Kreevoy and Oh assumed that there is no isotope effect on w^r or $\Delta G^{\circ\prime}$. The equation is useful only within the range $|\Delta G^{\circ\prime}| \leqslant \lambda$, and this prevents its use in consideration of many atom transfer isotope effects.

An alternative equation, also derived by Marcus[18], is from Johnston's[19] BEBO calculation of activation energies. It is specifically useful for these H atom transfer processes. This is Eq. (5), with the terms as above.

$$\Delta G^{\ddagger} = w^r + \frac{\lambda}{4} + \frac{\Delta G^{\circ\prime}}{2} + \frac{\lambda}{4 \ln 2} \ln \cosh \frac{2 \Delta G^{\circ\prime} \ln 2}{\lambda} \tag{5}$$

Using the same assumptions as Kreevoy and Oh, Eq. (6) can be written,

$$\Delta G_D^{\ddagger} - \Delta G_H^{\ddagger} = \frac{1}{4}(\lambda_D - \lambda_H) + (\lambda_D/4 \ln 2)\ln \cosh[(2 \Delta G^{\circ\prime}\ln 2)/\lambda_D] -$$

$$- (\lambda_H/4 \ln 2)\ln \cosh[(2 \Delta G^{\circ\prime}\ln 2)/\lambda_H] \tag{6}$$

an equation with a maximum at $\Delta G^{\circ\prime} = 0$, and becoming zero at extreme values of $\Delta H^{\circ\prime}$. The isotope effect is fairly readily calculated given λ_H, λ_D and $\Delta G^{\circ\prime}$, and $k_H/k_D = \exp[(\Delta G_D^{\ddagger} - \Delta G_H^{\ddagger})/RT]$. The equation is somewhat more difficult to handle analytically going from the isotope effect and $\Delta G^{\circ\prime}$ to λ or to find $\Delta G^{\circ\prime}$ from λ, although both are reasonably easy graphically.

An analytically simpler relation selected because of its simplicity and its correct limiting properties is the hyperbola, Eq. (7)[20] where $c = \Delta G^{\ddagger}$

$$\Delta G^{\ddagger} = \frac{\Delta G^{\circ} + \sqrt{\Delta G^{\circ 2} + 4 c^2}}{2} \tag{7}$$

when $\Delta G^{\circ} = 0$, analogous to Marcus' intrinsic barrier. Assuming that only c is affected by deuterium substitution, this is readily converted to Eq. (10).

$$k_H/k_D = \exp\left[\frac{1}{2RT}(\sqrt{\Delta G^{\circ 2} + 4 c_D^2} - \sqrt{\Delta G^{\circ 2} + 4 c_H^2})\right] \tag{8}$$

Equations such as (6) and (8) show that isotope effect maxima are a direct consequence of a non-linear relation between $\Delta G^{\ddagger}$ and ΔG°, such that $\Delta G^{\ddagger}$ approaches zero (or a constant value, w^r) as ΔG° becomes very negative, and approaches ΔG° when this becomes very positive[21].

These equations are not easily related to equations such as (2), since one proposes a mechanism (force constant asymmetry) for the fall of isotope effect, and the others ignore totally the isotope effect source, and merely assume that it exists, that is that $\lambda_D > \lambda_H$ or that $c_D > c_H$. It would of course be possible to calculate the term $\nu_{1H}^{\ddagger} - \nu_{1D}^{\ddagger}$ associated with any particular maximum isotope effect determined by $\nu_{HX} - \nu_{DX}$ from the isotope effect calculated with Eq. 6 (or Eq. 8), but the force constants could not be so calculated, since there is no unique set of force constants giving a particular $\nu_{1H}^{\ddagger} - \nu_{1D}^{\ddagger}$.

The Interpretation of Observed Isotope Effects

It is always tempting to try to fit observed quantities into a theoretical framework, no matter how simple and crude. One obstacle to such an attempt is the prevalence of experimental error. A discussion of experimental methods for the study of isotope effects in hydrogen atom transfers including some common errors, has been presented elsewhere[11] and will not be repeated. This reference also contains many of the experimental values on which the following discussion is based.

The discussion which follows is based principally on the interpretation in terms of Eqs. (6) or (8), and this is best introduced by a discussion of isotope effects in attack of carbon radicals on thiols. Several of these are presented in Table 1. In this table, and in the others later, isotope effects are presented as values of k_H/k_D at 25 °C. The published values are converted to 25 °C by the Eq. (9) where T is the experimental

$$(k_H/k_D)_{25°C} = (k_H/k_D)_T^{\frac{T}{298}} \qquad (9)$$

temperature in K except in cases where apparently reliable temperature dependence of the isotope effect was reported, then this dependence was used to calculate the 25 °C value. This correction using Eq. (9), may not be very good, but there is no better way to compare results at different temperatures. In many cases tritium rather than deuterium effects were used, in these the Swain-Schaad[22] Eq. (10), which is generally quite reliable[23], was used. The nature of the correction is indi-

$$k_H k_T = (k_H/k_D)^{1.442} \qquad (10)$$

cated in the fourth column in the Table 1. Many of the results are those of Pryor and Kneipp[24]. The parenthetical numbers refer to Fig. 1.

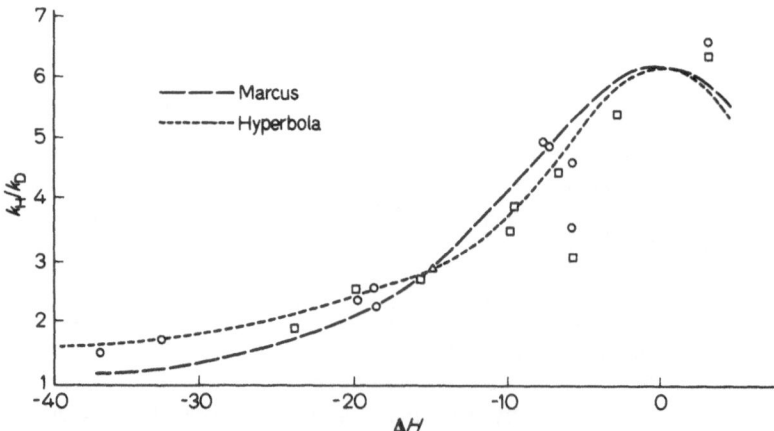

Fig. 1. Isotope effects (k_H/k_D for R· + HSR′ of different values of ΔH. □, from Ref.[24], ○ from Ref.[3] and [5]) of Table 1. The point △ and the maximum at $\Delta H = 0$ are the only parameters used to fit Eq. (6), ———, or Eq. (8) ...., to the data

Table 1. Isotope effects in R· + HSR′

R·	R′	Conditions[1]	k_H/k_D	
Ph· (3)	CMe_3	AT	1.9	[2]
p-$O_2NC_6H_4$ (4)	CMe_3	AT	2.5	[2]
CH_3· (8)	CMe_3	AD	2.7	[2]
1-Nonyl· (9)	CMe_3	AT	3.4	[2]
3-Heptyl· (11)	CMe_3	AT	4.4	[2]
Cyclohexyl· (14)	CMe_3	AT	3.0	[2]
Et_3C· (17)	CMe_3	AT	5.4	[2]
$PhCH_2$· (18)	CMe_3	AT	6.4	[2]
Ph_2CH·	CMe_3	AT	6.6	[2]
Ph_3C·	CMe_3	AT	6.7	[2]
1-Adamantyl·	CMe_3	AT	1.9	[2]
$PhCH_2$· (10)	Ph	AT	3.9	[2]
$PhCHCH_2SPh$ (12)	Ph	AT	4.8	[3]
m-$NO_2C_6H_4\dot{C}HCH_2SPh$	Ph	70T	5.4	[3]
$(p$-$ClC_6H_4)_2\dot{C}CH_2SPh$	Ph	70T	5.7	[3]
$(pMeOC_6H_4)_2\dot{C}CH_2SPh$	Ph	70T	8.3	[3]
$AcO\dot{C}HCH_2SPh$	Ph	70T	2.2	[3]
$PhCH_2\dot{C}HCH_2SPh$ (6)	Ph	70T	2.2	[3]
n-$C_6H_{13}\dot{C}HCH_2SPh$ (7)	Ph	70T	2.6	[3]
n-$C_6H_{13}\dot{C}HCH_2SCH_2Ph$ (16)	$PhCH_2-$	70T	4.6	[3]
$PhCH_2CHCH_2SCH_2Ph$ (15)	$PhCH_2-$	70T	3.5	[3]
$AcO\dot{C}HCH_2SCH_2Ph$ (15)	$PhCH_2-$	70T	4.0	[3]
p-$MeOC_6H_4\dot{C}HCH_2SCH_2Ph$	$PhCH_2-$	70T	6.0	[3]
$Ph\dot{C}HCH_2SCH_2Ph$ (19)	$PhCH_2-$	70T	6.5	[3]
m-$O_2NC_6H_4CHCH_2SCH_2Ph$	$PhCH_2-$	70T	6.9	[3]
$(p$-$ClC_0H_4)_2\dot{C}CH_2SCH_2Ph$	$PhCH_2-$	70T	5.3	[3]
$PhCHCH_2SMesityl$ (13)	Mesityl	70T	4.8	[3]
Ph_3C·	Mesityl	AT	9.4	[3]
Ph_3C·	Ph	AT	6.5	[4]
p-$NO_2C_6H_4$· (2)	Ph	60T	1.7	[5]
p-ClC_6H_4	Ph	60T	1.6	[5]
·Ph· (1)	Ph	60T	1.5	[5]
α-$C_{10}H_7$·[6]	Ph	60T	1.3	[5]
p-$NO_2C_6H_4$	p-ClC_6H_4	60T	1.8	[5]
Ph·	p-ClC_6H_4	60T	1.6	[5]
α-$C_{10}H_7$·[6]	p-ClC_6H_4	60T	1.5	[5]
p-$NO_2C_6H_4$·	CMe_3	60T	2.4	[5]
p-ClC_6H_4	CMe_3	60T	2.4	[5]
α-$C_{10}H_7$·[6]	CMe_3	60T	2.2	[5]

[1] The letter A shows that an observed Arrhenius temperature dependence of isotope effect was used in the reduction to 25 °C; a number is the temperature of measurement in °C with reduction to 25° by Eq. (9). T denotes use of tritium and Eq. (10).

[2] Ref.[24].

[3] Lewis, E. S., Butler, M. M.: J. Am. Chem. Soc. 98, 2257, 1976.

[4] Lewis, E. S., Butler, M. M.: J. Org. Chem. 36, 2582 (1971).

[5] Lewis, E. S., Ogino, K.: J. Am. Chem. Soc. 98, 2260 (1976).

[6] α-Naphthyl.

Using an average SH stretching frequency of 2570 cm^{-1}, $(k_H/k_D)_{max}$ is 6.15 at 25 °C. Only two of the entries in the table are substantially in excess of this, the reaction of the substituted 1,1-di-p-anisylethyl radical with thiophenol, and the reaction of the trityl radical with mesitylene thiol, and in this latter case the large temperature dependence and small A_H/A_D factor suggested a substantial tunnel correction. Only one case appears twice ($pNO_2C_6H_4$ with t-butyl mercaptan with k_H/k_D given as 2.5 and 2.4), and the agreement is satisfactory, but Pryor and Kneipp's value for phenyl radicals would not fit as well into the series of Lewis and Ogino. An estimate of $\pm 10\%$ uncertainty would probably take care of most discrepancies.

The table is incomplete, it does not contain even all the results on transfer from sulfur to carbon in the papers quoted. There are several entries showing substituent effects, which are supported by more examples in the original. The examples of H_2S as a hydrogen atom donor are omitted because of the inapplicability of the 1-dimensional treatment. Because of serious failure of the assumption that A and X are heavier than H, the cases of A· = H· have been omitted.

In Fig. 1 are plotted values of k_H/k_D vs. ΔH, as originally done by Pryor and Kneipp. The points are all taken from the table and include all those where a reasonable estimate of ΔH could be made from bond energies. The points are identified with the numbers in parentheses in Table 1. The points originally included by Pryor and Kneipp that have been excluded are those of attack on H_2S, attack by H·, attack by diphenylpicrylhydrazyl (on the grounds that it is not a carbon radical) and the points for benzhydryl, trityl, and adamantyl radicals on the grounds that D_{RH} is inadequately known. Several points from the other work are included; in the case of radicals with β sulfur, it was assumed that these did not affect the bond dissociation energy. The fact that these points in fact fall rather close to the curve and show no systematic deviations can alternatively be used to confirm the energetic insignificance of participation by neighboring sulfur.

Several points for the series $XMe_2C·$ + HSR have been omitted. The isotope effects were experimentally difficult k_D/k_T measurements, and the errors in converting these to k_H/k_D may be large.

The resulting plot, Fig. 1, can clearly not be adequately fitted by any smooth curve, nevertheless a trend is obvious. The figure shows a curve plotted according to Eq. (6) and another according to Eq. (8). The assumption has been made that $\Delta G°' = \Delta G° = \Delta H°$, which may not be very wrong. Both have maximum isotope effects of 6.1, and both have been adjusted to go through the point $k_H/k_D = 2.85$ at $\Delta H = -15$ kcal and these are the only adjustable parameters. It can be seen that there is not much to choose between them, and that they both describe the obvious trend adequately, but with several serious failures. Since the values of ΔH may be uncertain by a couple of kilocalories, the fits to the curves are almost as good as can be expected. However, the isotope effects for the benzhydryl and trityl systems are almost certainly too high to fit on the curves for any reasonable values of D_{RH}. These cases of rather large isotope effects may be thrown off by modest tunnel correactions, which are not easily demonstrated.

If we accept the curves as representing the trend, we can get the intrinsic barrier. The Marcus curve drawn has $\lambda/4 = 5.10$ Kcal/mole, the "hyperbolic" curve has $c = 4.73$ Kcal/mole, again in satisfactory agreement. The apparent displacement of the maximum from $\Delta H° = 0$ may be a distortion of the curve by tunnel corrections, or may be real, in which case the problem could lie with the various crude approximations, such as $\Delta H° = \Delta G°$, $\Delta G° = \Delta G°'$, and the absence of isotope effect in $\Delta G°$ or w^r. The attempt to fit makes the implicit assumption that λ (or c) is constant through the series. We assume tentatively that in the absence of steric effects λ and c are determined only by the nature of the atoms to which H is directly attached in reagent and product. It has already been shown[25] that for equal exothermicity, the isotope effect is highly variable as these atoms are varied, implying variable λ and c, and therefore preventing any far more general constancy assumption.

If these assumptions of the approximate fit of Eqs. (16) or (8), and the derivation of the intrinsic barrier are reasonable, further conclusions can be drawn. Consider, for example, the reaction of trityl radical with thiophenol. This has an activation energy of about 9.1 Kcal/mole, as determined directly by Glaspie[26]. Using the enthalpy form of the hyperbolic relation $E_a(E_a - \Delta H) = c^2$, we find using $c = 4.73$, $\Delta H = +6.6$ Kcal., and using $D_{SH} = 75$ Kcal/mole[35], we then find $D_{Ph_3C-H} = 68$ Kcal/mole. Although this value is very rough, it is for example, compatible with the great instability of $Ph_3CN_2CPh_3$ shown by the failure to isolate it[27]. One could, as originally suggested[28], determine many other bond dissociation energies, with, however, the restriction that only isotope effects (k_H/k_D at 25°) between 2.5 and 5 would be particularly useful. This restriction is advisable because the very low slope of the curve (as well as the choice of function) makes ΔH too sensitive to the isotope effect below a value of about 2, and the values much above 5 may have significant tunnel corrections.

Using these methods, it will be of interest to look at a number of other atom transfer isotope effects. Table 2 contains examples of attack on CH bonds by various first-row radicals, mostly from the compilation by Russell[29].

Several other examples from the references quoted are omitted, and a major omission is the thorough study of the reactions $CF_3· + HCD_3$ and $CD_3· + C_2H_6$ over

Table 2. Isotope effects in first-row X· + RH reactions

X·	R	Conditions[1]	k_H/k_D, 25°	
$Me_3CO·$	$H-CH_2C_6H_4Cl$-m	40 T	7.4	[2]
$Me_3CO·$	$H-CH_2Ph$	40 T, D	6.2	[2]
$PhMe_2O_2$	$H-CMe_2PH$	30 D	8.3	[3]
$PhMe_2O_2$	$H-CMe_2Ph$	60 D	6.7	[3]
Me_3CO_2	$HCMe_2Ph$	30 D	21.0	[3]
CH_3	HCH_2Ph	164 D	21.0	[3]
Ph·	$H-CH_2Ph$	60 D	5.0	[3]
$Me_2N·$	$H-CH_2Ph$	136 D	6.7	[3]
Ph	HCH_2COCH_3	60 D	5.0	[3]

[1]) The number indicates the temperature at which the measurement was made, T shows the use of tritium, D, deuterium, T, D shows that both were used without much difference.
[2]) Ref.[25].
[3]) Ref.[29].

a very wide temperature range in the gas phase by Johnston in Ref.[2], where a thorough treatment of the substantial tunnel correction as well as the transition state vibrations is considered. Although some of these reactions are exothermic by as much as 20 Kcal, the isotope effects are a large fraction of the maximum, 7.8, suggesting a large value for the intrinsic barrier. The barrier for the symmetrical $CH_3 \cdot + CH_4$ reaction is indeed about 14 Kcal/mole, and for the very unsymmetrical $(CH_3)_3CO \cdot + H_3CPh$ reaction, exothermic by 19 Kcal, the activation energy is 5.6 ± 2 kcal[30], corresponding to a value of c of 11.7 Kcal/mole by the hyperbolic treatment. Analogously, the thermoneutral $CH_4 + \cdot OCH_3$ reaction has $E_a = 11$ Kcal/mole[31]. We can reword the earlier conclusion to state that transfers between first-row radicals will have a large value of c. The major exception to this conclusion, for which there is no obvious explanation, is that the isotope effect in the transfer of hydrogen to radicals in polymerizing vinyl acetate has $k_H/k_D = 1$, when the source is m-cyanophenol, although some other phenols give isotope effects at least as large as the "maximum" of 13[32].

The transfers involving halogen atoms are of traditional interest. Some of these are shown in Table 3, together with some results on the reverse reaction of radicals with HBr. They are again reduced for comparison purposes to k_H/k_D at 25°. The

Table 3. Isotope effects involving halogen atoms

Radical	H Atom Donor	Cond.[1]	k_H/k_D	
Cl·	H–CH$_3$	0, D	9.8	[2]
Cl·	H–CH$_2$CH$_3$	127 D	3.4	[2]
Cl·	H–C(Me)$_3$	−15 D	1.3	[2]
Cl·	H–CCl$_3$	127 D	2.5	[2]
Cl·	H–CH$_2$Ph	40, D, T	2.2[3]	[4]
Br·	H–CH$_2$Ph	40, D, T	6.8[3]	[4]
Br·	H–CHMePh	77 D	3.2	[5]
Br·	H–CMe$_2$Ph	77 D	2.0	[5]
Br·	H–CMe$_3$	0, T	2.2	[6, 7]
Br·	H–CMe$_2$CH$_2$CH$_3$	0, T	2.4	[6, 7]
Br·	H–CMe$_2$CH$_2$Cl	0, T	3.8	[6]
Br·	H–CMe$_2$CH$_2$Br	0, T	4.6	[6]
p-CH$_3$C$_6$H$_4$ĊHCH$_2$Br	H–Br	0, T	1.03	[8]
PhĊHCH$_2$Br	H–Br	0, T	1.27	[8]
p-ClĊHCH$_2$Br	H–Br	0, T	1.32	[8]
n-C$_6$H$_{13}$ĊHCH$_2$Br	H–Br	0, T	1.59	[8]
BrĊHCH$_2$Br	H–Br	0, T	1.94	[8]
·CH$_2$CH$_2$Br	H–Br	0, T	2.49	[8]

[1] As in Tables 1 and 2.
[2] From Ref.[29].
[3] An average of several determinations.
[4] Ref.[25].
[5] Wiberg, K. B., Slaugh, L. H.: J. Am. Chem. Soc. 80, 3033 (1958).
[6] Ref.[6b].
[7] This is close to the equilibrium isotope effect.
[8] Ref.[6a].

reaction of chlorine atoms are exothermic, with the exception of the nearly thermo-neutral reaction with methane. The reaction with methane has an activation energy of about 3.8 Kcal/mole[31], so that the fast drop off from the maximum of about 7 is to be expected, thus Eq. (10) gives $k_H/k_D = 2.3$ for $c = 3.8$, $\Delta H = 19$ Kcal. We shall comment later on the smaller value of c for the Cl–H–C system than the S–H–C system.

The systems involving bromine atoms also show the fast drop-off as ΔH deviates from 0, especially marked in the toluene to cumene series, which implies a small value of the intrinsic barrier. There is a further problem that there is a substantial equilibrium isotope effect of about a factor of 2^{6b}, accompanied in the aliphatic cases by considerable opportunity for reversibility, thus k_H/k_D values in the endothermic attack of Br· on aliphatic RH are unlikely to be less than 2. Correspondingly, isotope effects in the reverse directions can be inverse. This has not been observed, but the trend shown in the last several reactions, in which isotope effects as low as 1.0 are observed suggests this possibility. We attach no meaning to $(k_H/k_D)_{max}$ for abstraction from HBr because of the failure of the one-dimensional model. Both chlorine and bromine then fit the scheme of highly variable isotope effects associated with low intrinsic barriers.

Two other systematic solution phase tritium isotope effects studies have been made that will only be mentioned here. In the reduction of alkyl halides by trialkyl-tin hydrides, there is an abstraction step

$$R\cdot + HSnR'_3 \rightarrow RH + \cdot SnR'_3$$

Isotope effects are small, $(k_H/k_D)_{max}$ at 25° is only 3.6 and values varying from 2.2 to 3.1 were found[33]. The correlation with bond energy only showed up in the contrast between the aliphatic and benzyl radicals. Another study[34] was the addition of dimethyl phosphonate and dimethyl thionephosphonate to olefins, in which the transfer step is reaction (11). Here variation of R with both X=O and

$$R\overset{\cdot}{\underset{H}{C}}-CH_2\overset{X}{P}(OMe)_2 + H-\overset{X}{P}(OMe)_2 \rightarrow RCH_2CH_2\overset{X}{P}(OMe)_2 + XD(OMe)_2 \qquad (11)$$

X=S gave isotope effects mostly as large as or larger than the $(k_H/k_D)_{max}$ of about 5.6 at 25°. Steric hindrance was believed to slow down the rates, increase the classical isotope effect, and increase the tunnel correction.

Conclusions

The study of isotope effects in hydrogen transfer reactions is an experimentally accessible and not difficult approach to the study of the transition state for these reactions. In series of uncomplicated systems characterized by a constant intrinsic barrier, the relation between the barrier weight, the isotope effect, and the exo-thermicity allows the calculation of any one of these from the others using the

BEBO-based Marcus equation or the hyperbolic relation which is not based on theory but is easier to use and gives about the same answer. The reduction in isotope effect predicted by the Westheimer transition state "symmetry" arguments is harder to correlate, yet the wide range of isotope effects virtually requires this explanation, in contrast to the case in proton transfers.

We can also consider cases in which the intrinsic barrier is altered. Two such effects are steric hindrance and contribution of charge-separated structures to the transition state. Steric hindrance raises the energy of the transition state compared to that of a similarly exothermic unhindered model. This can be accomodated by considering an increase in the intrinsic barrier, which therefore makes the isotope effect rise. In ref.[11] this is alternatively interpreted in a quadratic representation of the surface as an increase in the interaction force constant, and thus also correlated with an increase in the tunnel correction. An example of such an enhancement is the large value of the isotope effect in the trityl radical mesitylenethiol reaction in Table 1.

The contribution of polar structures reduces the barrier and also the intrinsic barrier. This results for non thermoneutral reaction in a reduction of isotope effect. This has been a controversial subject for several years; it is extensively covered by Russell[29]. The variation with substituents in the low isotope effects for the reaction of aryl radical with arene thiols were explained using such an effect. We may possibly further account for the lower intrinsic barrier for the R—H—Cl system (3.8 Kcal) than for the R—H—S system (5 Kcal) in terms of the greater electronegativity of chlorine.

A measure of transition state symmetry was proposed earlier[11, 6b], based on a simple way of drawing energy-reaction coordinate diagrams. This measure p (called f in Ref.[11]), and named the fraction of product-like character, is defined by Eq. (12).

$$p = \frac{E_a}{2\,E_a - \Delta H^\circ}, \text{ or } \frac{\Delta G^\ddagger}{2\,\Delta G^\ddagger \Delta G_0} \tag{12}$$

It is of some interest to relate the isotope effect to p, using Eq. (6). This is illustrated in Fig. 2, using a plot that, to the extent that Eqs. (9) and (10) are correct is applicable to either k_H/k_D or k_H/k_T measurements at any temperature. The curve shown uses

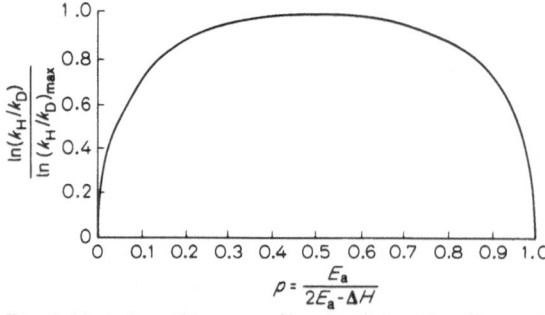

Fig. 2. Variation of isotope effect, calculated by Eq. (6), with the symmetry measure $p = E_a/(2\,E_a - \Delta H)$. The curve is calculated for the same value of λ_H and λ_D that were used in Fig. 1, but this curve is insensitive to this choice, except for very small values of λ_H.

p calculated with an intrinsic barrier of 5.1 Kcal for the H compound, and $(k_H/k_D)_{max} = 6.1$ (as in Fig. 1). The curve is dependent on the intrinsic barrier, but with a barrier twice as great, the deviation is too small to show on the graph. Very small barriers (where p is significantly different for H and D) deviate more strongly, but the form is very similar. Thus, should the necessary data be available, p is an entirely satisfactory measure of T.S. symmetry, and a curve such as Fig. 2 could be generally useful.

Acknowledgement. Most of the work from Rice on which this paper is based was supported by a grant from the Robert A. Welch Foundation, which is gratefully acknowledged.

References

1) Recent studies include:
   a) Bell, R. P.: The proton in chemistry. 2nd Ed. New York: Cornell University Press 1973
   b) Caldin, E. F., Gold, V., (eds.): Proton transfer reactions. London: Chapman and Hall 1975
   c) Proton transfer: Faraday Symposium of the Chemical Society, No. 10 (1975)
2) A highly relevant discussion of gas-phase H atom transfer isotope effects is Johnston, H. S.: Gas phase reaction rate theory. Chap. 13. New York: Ronald Press 1966
3) Biegeleisen, J.: J. Chem. Phys. *17*, 675 (1949)
4) Theoretical evidence for linearity of H atom transfer transition states, Ref.[2], Chap. 10. The evidence for linearity of proton transfer transition states is based upon analogy to the inverting S_N2 reactions on carbon, the linearity of HF_2^- etc., as well as electrostatic and orbital considerations
5) Lewis, E. S., Symons, M. C. R.: Quart. Revs. *12*, 230 (1958); Hawthorne, M. F., Lewis, E. S.: J. Am. Chem. Soc. *80*, 4296 (1958); Olah, G. A., Halpern, Y., Shen, J., Mo, Y. K.: J. Am. Chem. Soc. *93*, 1251 (1971)
6a) Lewis, E. S., Kozuka, S., among others: J. Am. Chem. Soc. *95*, 282 (1973)
   b) Lewis, E. S., Shen, C. C.: J. Am. Chem. Soc. *99*, 3055 (1977)
7) Westheimer, F. H.: Chem. Rev. *61*, 265 (1961)
8) We shall use highly exothermic reactions with reagent-like transition states throughout as models for unsymmetrical reactions, rather than always adding "or endothermic reactions with product-like transition states". This avoids verbiage and problems associated with the usually small equilibrium isotope effects. It is also experimentally reasonable, for free radicals in organic systems seldom stay around long enough to react in highly endothermic reactions. Such reactions can of course be studied in the reverse direction
9) Bell, R. P.: The proton in chemistry. 2nd edit. New York: Cornell University Press 1973, p. 267 ff.
10) Willi, A. V.: Helv. Chim. Acta *34*, 1220 (1971)
11) Lewis, E. S., Chap. 4 in: Isotopes in hydrogen transfer processes. Vol. 2 of: Isotopes in organic chemistry. Buncel, E., Lee, C. C., (eds.). Amsterdam: Elsevier 1976
12) Lewis, E. S.: Chap. 10 in Ref.[1b]
13) More O'Ferrall, R. A.: p. 238, Ref.[1b]
14) We ignore the case that $\nu_1^{\ddagger}H > \nu_{HX}$; it seems unrealistic
15) An example in which a k_H/k_D of 10 is divided about equally between the tunnel correction and the semiclassical isotope effect for a hydride transfer is given in: Lewis, E. S., Robinson, J. K.: J. Am. Chem. Soc. *90*, 4337 (1968)
16) Marcus, R. A.: J. Chem. Phys. *24*, 966 (1956)
17) Kreevoy, M. M., Oh, S.: J. Am. Chem. Soc. *95*, 4805 (1973)
18) Marcus, R. A.: J. Phys. Chem. *72*, 891 (1968); Marcus, R. A., Chapt. 1 in: Technique of chemistry. Vol. VI. Investigations of rates and mechanisms of reactions. Lewis, E. S. (ed.). New York: Wiley-Interscience 1974

19) Ref.[2] Chap. 10
20) Lewis, E. S., More O'Ferrall, R. A., Shen, C. C.: in preparation; Chu, C. C.: Ph. D. Thesis, Rice University, 1976; Lewis, E. S., More O'Ferrall, R. A.: paper presented at A.C.S. SW Regional Meeting, El Paso, 1973
21) This relation between variable isotope effects with a maximum and plausibly non-linear free energy relation was qualitatively noted in connection with proton transfers: Lewis, E. S., Funderburk, L. H.: J. Am. Chem. Soc. 89, 2322 (1967). Marcus[18] correctly points out that the reference to diffusion control is irrelevant
22) Swain, C. G., Stivers, E. C., Renwer, J. F., Schaad, L. J.: J. Am. Chem. Soc. 80, 5885 (1955)
23) Lewis, E. S., Robinson, J. K.: J. Am. Chem. Soc. 90, 4337 (1968); Jones, J. R.: Trans. Faraday Soc. 65, 2138 (1969)
24) Pryor, W. A., Kneipp, K. G.: J. Am. Chem. Soc. 93, 5584 (1971)
25) Lewis, E. S., Ogino, K.: J. Am. Chem. Soc. 98, 2260 (1976)
26) Glaspie, P. S.: Ph. D. Thesis, Rice University, 1974
27) Wieland, H.: Ber. 42, 3020 (1909); Pinck, L. A.: J. Am. Chem. Soc. 55, 1711 (1933)
28) Lewis, E. S., Butler, M. M.: Chem. Comm. 1971, 941
29) Russell, G. A., in: Free radicals. Kochi, J. K. (ed.). New York: Wiley-Interscience 1973, p. 312
30) Carlsson, D. J., Howard, J. A., Ingold, K. U.: J. Am. Chem. Soc. 88, 4725 (1966)
31) Kerr, J. A., in: Free radicals. Vol. I. Kochi, J. K. (ed.). New York: Wiley-Interscience 1973, p. 15
32) Quoted by Simonyi, M., Tüdos, F.: Adv. Phys. Org. Chem. 9, 127 (1970)
33) Kozuka, S., Lewis, E. S.: J. Am. Chem. Soc. 98, 2254 (1976)
34) Lewis, E. S., Nieh, E. C.: J. Am. Chem. Soc. 98, 2268 (1976)
35) Added in proof: Values of S—H bond energies have recently been revised upward by more than 5 Kcal/mole. If all else is unchanged, the estimate of the trityl-H bond energy should be correspondingly increased

Received July 20, 1977

N-Methylacetamide as a Solvent

Dr. Robert J. Lemire and Prof. Dr. Paul G. Sears

Department of Chemistry, University of Kentucky, Lexington, KY 40506, U.S.A.

Table of Contents

I. Introduction

Interest in the N-methylamide derivatives of the lower carboxylic acids as potential solvents was triggered by the reporting of the exceptionally high dielectric constants of these liquids by Leader and Gormley[1] in 1951. N-methylformamide (NMF), N-methylacetamide (NMA), N-methylpropionamide (NMP) and, to a much lesser extent, N-methylbutyramide (NMB) subsequently were investigated as new non-aqueous solvents. During the twenty-six years since Leader and Gormley initially reported the unique dielectric properties of some of the N-methyl amides, NMA has been investigated by far the most extensivley among these compounds. Its selective and broader use can, in major part, be attributed not only to its relatively greater ease of purification but also to its best overall combination of properties.

Previously, NMA as a solvent has been discussed briefly in books by Charlot and Trémillon[2] in 1963 and Waddington[3] in 1965 and reviewed fairly extensively by Dawson[4] in 1963 and Vaughn[5] in 1967 (although this latter review contains no references beyond 1963). Electrochemistry in NMA was surveyed by Reid and Vincent[6] in 1968 and purification and tests for purity of NMA have been discussed in detail by Knecht[7] in 1971.

In view of the already extensive and rapidly expanding literature concerning NMA, the present comprehensive review will be restricted to the use of pure NMA as a solvent. The literature has been covered rigorously through 1975 and in major part through 1976. Much of the NMA-related literature, especially that in the 1970's, deals with the properties and chemistry of mixed-solvent systems in which NMA is combined with water, N,N-dimethylformamide, N,N-dimethylacetamide, t-butyl alcohol, dimethyl sulfoxide, or other solvents; hence, it is currently anticipated that the broad topic of NMA-containing mixed-solvent systems will be covered in another review.

II. Synthesis and Purification

a) Synthesis

As commercially available NMA is relatively expensive and only moderately pure, it is often convenient to prepare this amide in the laboratory when carrying out studies involving large quantities of the solvent. Several similar synthetic methods have been described which involve the addition of methylamine gas[5] or of methylamine in concentrated aqueous solution[8] to acetic acid. The addition leads to the formation of methylammonium acetate which then must be heated to crack out and distill off water yielding crude NMA. The optimum temperature to which methylammonium acetate should be heated to facilitate cracking, while minimizing subsequent decomposition of the NMA product, has not been established. It appears, however, that heating the mixture above 180 to 190 °C results in a markedly poorer crude amide product and leads to increased difficulty with the subsequent purification.

A second convenient general preparative method involves the addition of methyl-amine to methyl or ethyl acetate[9, 10]. The mixture is allowed to stand for approximately two weeks under anhydrous conditions. This procedure yields crude NMA after removal of the alcohol by-product and the unreacted starting materials by distillation under vacuum.

b) Purification

Initial purification of NMA can be effected by several cycles of fractional freezing[11] (initially at temperatures as low as −20 °C), by distillation under vacuum[7] (it has been suggested[12] that addition of small quantities of an oxidizing agent aids purification), or by a combination of these methods[5]. Distilling the crude material at atmospheric pressure and retaining the fraction with a boiling point greater than 204 °C will yield a somewhat poorer product. By these methods, NMA can be obtained which melts between 28 and 30 °C; this is not sufficiently pure, however, for several types of investigations such as polarographic studies[7].

Further purification can involve tedious methods incorporating extraction and treatment with acidic and basic reagents and also treatment with drying agents[7]. For reasonably small quantities of very highly purified solvent (up to about one liter), zone refining of NMA has been recommended[13].

When partially purified NMA is subjected to several (5−15) cycles of fractional freezing under a nitrogen atmosphere (NMA is hygroscopic[5]), a product can be obtained which has a melting point comparable to the zone-refined material[13, 14] and which is free of impurities detectable by polarography[15, 16]. The fractional freezing method has the advantage that larger quantities of the solvent (4 to 5 liters) can be handled at one time.

The pure solvent has been found to decompose slightly — probably to an equilibrium mixture containing small quantities of methylamine and N,N-diacetylmethyl-amine[17]. This decomposition is accelerated by contact with a metal surface[17]. The formation of such decomposition products might be the primary cause of the problems involved in some of the purification procedures[7]. Decomposition apparently occurs slowly even in solid NMA.

III. Structure of NMA Other Than in the Liquid Phase

a) Structure and Dipole Moment of NMA in the Gas Phase and in Dilute Solution

Studies of NMA in the gas phase and in dilute solution in a variety of solvents have been used to determine information concerning the NMA molecule and its hydrogen bonding interactions. In very dilute solution in a non-polar solvent, it would be expected that NMA would behave structurally in a very similar manner to NMA in the gas phase.

The structure of NMA in the gas phase was partially determined by Kimura and Aoki[18] using electron diffraction techniques. Recently, Kitano et al.[19] have re-investigated the structure in more detail (Table III-1). Analysis of the data suggested that conformers other than that with a *trans* configuration (about the C–N bond) represent less than 5% of the molecules. The mass spectrum of NMA has been reported by Gilpin[20] and by Tereshkovich and co-workers[21] and the first (vertical) ionization potential of NMA has been found to be 10.21 eV[22].

Table III-1. Structural parameters for N-methylacetamide in the gas phase and in the crystalline phase (−35 °C)

C_1 C_2 N C_3 (with $\parallel$ O on C_2)	Electron diffraction (gas) Ref.[19]	X-Ray (crystal, −35 °C) Ref.[60]
C_1–C_2	1.52_0 A	1.53_6 A
C_2–N	1.38_6	1.29_0
C_2–O	1.22_5	1.23_6
N–C_3	1.46_9	1.46_5
C–H (avg)	1.10_7	
$\not\prec$ N–C_2–O	121.8°	123°
$\not\prec$ C_2–N–C_3	119.7°	120.5°
$\not\prec$ C_1–C_2–N	114°	116.5°
$\not\prec$ C_1–C_2–O		120.5°
$\not\prec$ H–C–H (avg)	110°	
$\not\prec$ C_2–N–H	110°	

The dipole moment of NMA has been calculated from vapor phase measurements[23] to be 3.71 D. This compares reasonably well with 3.82 D as determined from dilute solutions of NMA in benzene[23] and with 3.6 D as determined from dilute solutions in carbon tetrachloride[24]. These values are somewhat lower than the earlier values obtained from more concentrated solutions of NMA in dioxane and carbon tetrachloride by Mizushima et al.[25]. The vector moment of NMA is 3.6 D[26] based on the molecular structure determined by electron diffraction[18].

The vacuum UV spectrum of gaseous NMA has been examined as a function of pressure[27, 28]. It was noted that as the pressure was increased one set of absorption bands diminished in intensity while a new set of bands appeared at slightly higher wavelengths and increased in intensity. This was attributed to dimer formation in the gas phase. Nielsen and Schellman[29], however, found little change in the UV spectrum of NMA on dilution of a solution in cyclohexane to 10^{-3} M. They concluded that the formation of dimers and larger aggregates of NMA had little effect on the UV spectrum of NMA.

Miyazawa, Shimanouchi and Mizushima[9], Jones[30] and Hallam and Jones[31] examined the vapor phase infrared spectrum of NMA but it has not been possible to deduce the predominant conformation of the molecule from the available data (although N-methylformamide was concluded to exist primarily as the *trans* conformer[31])). The infrared spectrum of dilute solutions of NMA in CCl_4 has been examined by a large number of workers. It has been noted[32-34] that bands for both the *cis* and *trans* forms of the amide are found with only about 5% of the NMA being in the *cis* configuration[33]. Evidence that the *trans* configuration is predominant also has been found from the proton NMR spectrum with suggests 3 to 7% of the *cis* form in aqueous solutions of NMA[35, 36] but somewhat lesser percentages of the *cis* form for NMA dissolved in several other solvents[36].

It has been shown that hydrogen bonding between NMA molecules is quite extensive even in fairly dilute solution in non-polar solvents. Klotz and Franzen[37] determined a dimerization constant of 4.7 for NMA in CCl_4 using infrared techniques. Solvents which could participate in hydrogen bonding with NMA molecules such as dioxane and water showed markedly lower NMA dimerization constants. Graham and Chang[38] used NMR spectroscopy to estimate the enthalpy of hydrogen bonding of NMA in CCl_4 solutions to be -15 kJ mol^{-1}. Many similar studies have been done using a wide variety of techniques[39-46]. Although agreement between different workers is not particularly good[38], association constants for association beyond the dimer are apparently an order of magnitude larger than the dimerization constants. Association of N-deuterio-N-methylacetamide also has been examined in dilute CCl_4 solutions[46, 47]. Lǿwenstein and co-workers[46] found that the dimerization constant for the N-deuterio compound was slightly higher than for N-protio NMA while the K_A's for polymer formation were somewhat lower. Kresheck et al.[47], however, suggest that the deuterated compound forms longer polymeric chains than does ordinary NMA.

Several molecular orbital treatments of NMA have been reported[48]. Most of these are semi-empirical (*e.g.*, CNDO/2[49-51] or EHT[50, 51]) but Shipman and Christofferson have reported[48] an *ab initio* study which gives reasonable agreement between the calculated and experimental dipole moments and which predicts the difference in energy between the *cis* and *trans* configurations to be 15.3 kJ mol^{-1} (*i.e.*, >99% *trans* at 25 °C).

Perricaudet and Pullman[52] have also done an *ab initio* study which suggests that a small amount of twisting about the C—N bond (up to about 15°) can occur with very little loss of energy. It should be noted that the structural parameters in both of these studies were based on an average peptide geometry, not on experimental structural parameters determined particularly for NMA.

Lakshminarayanan and co-workers have suggested on the basis of semi-empirical molecular orbital studies[49, 53] that the methyl group and the proton attached to the nitrogen may be slightly out of the $CH_3(O)CN$ plane in the minimum energy configuration but that the actual energy difference between the planar and nonplanar configurations is very slight. Such a distortion has also been suggested by the NMR spectrum of NMA oriented in the nemantic phase of a liquid crystal[54]. A more marked but similar non-planarity has been found experimentally for formamide[55, 56].

Free energies of activation for rotation about the C—N bond have been determined by analysis of NMR lineshapes by Drakenberg and Forsén[57] to be 87 kJ · mol^{-1} (35% NMA in 1,2-dichloroethane) and 89 kJ · mol^{-1} (20% NMA in H_2O) at 60 °C. The difference between these values is probably not significant. Unfortunately, at these concentrations in these solvents, solute-solute and solvent-solute interactions would be expected to be such that the experimental barrier to rotation probably differs slightly from both the barrier to rotation for pure liquid NMA and from that for isolated NMA molecules. The values are close to those for other amides (*e.g.*, $\Delta G^*_{128} = 88$ kJ $\cdot$ mol^{-1} for N,N-dimethylformamide)[58]. Similar barriers have been predicted theoretically[52, 59].

b) Structure of Solid NMA

NMA has been found to have a solid phase transition at about 10 °C and the detailed crystal structure of the lower temperature form has been determined by X-ray diffraction[60]. The reported intramolecular bond lengths and angles are listed in Table III-1. The main difference between the gas phase and crystal structures is found in the length of the C^2–N bond which is markedly longer in the gas phase NMA molecule. This difference has been attributed to the intermolecular hydrogen bonding in the solid[19]. The NMA molecules are arranged in layers (separation 3.26 Å) and, within each layer, molecules are linked by hydrogen bonds (2.83 Å). The higher temperature solid form has two molecules per unit cell ($a = 4.85$ Å, $b = 6.59$ Å, $c = 7.30$ Å) compared to the four molecules per unit cell in the lower temperature form ($a = 9.61$ Å, $b = 6.52$ Å, $c = 7.24$ Å). Katz und Post[60] suggested that the NMA molecule can be hydrogen bonded in two different orientations and that above 10 °C the actual orientation occurring may be random. Dentini *et al.*[61] have discussed the two different forms in terms of van der Waals forces and hydrogen bonding. Kojima and Kawabe[62] have studied the dielectric properties of solid NMA and confirmed the existence of the solid phase transition at about 10 °C (9.3 ± 0.4 °C). The complex part of the dielectric constant increases sharply at this temperature and at about the same temperature a plot of resistivity versus temperature suddenly decreases. These workers consider that the dielectric constant anomaly at the transition point is "due to the rotational motion of the amino group". Suga *et al.*[63] used differential thermal analysis to estimate the heat of the solid phase transition ($\approx 4.2 \times 10^2$ J $\cdot$ mol^{-1} at ≈ 2 °C).

Bradbury and Elliot have reported[64] the infrared spectrum of crystalline NMA. Polarized spectra were run with the E vector parallel to each of the crystal axes in turn. The temperature dependence of the spectra was quite marked (see also Ref.[9]) and this was attributed, at least in part, to the previously mentioned solid phase transition. Apparently the atomic motions in NMA are of considerable complexity. The far-infrared spectrum has been reported by Itoh and Shimanouchi[65]. Schneider and co-workers[66–68] have done an extensive study of the vibrational spectra of solid NMA and its deuterated analogues and have done complete normal coordinate analyses of these compounds[69].

Aihara[70] has measured the vapor pressure of pure solid NMA over the temperature range of 15 to 30 °C. From these data he has calculated values (298.1 K) of the free energy (20.6 kJ $\cdot$ mol^{-1}), enthalpy (54.1 kJ $\cdot$ mol^{-1}) and entropy (112 J $\cdot$ mol$^{-1} \cdot$ K^{-1}) of sublimation for NMA. It should, however, be noted that the value determined for the enthalpy of sublimation at 298.1 K is slightly smaller than the calculated value of the enthalpy of vaporization of liquid NMA at 100 °C (cf. Section IV-b).

c) Structure of Matrix Isolated NMA

The structure of NMA isolated in nitrogen and argon matrices has been studied by Fillaux and de Lozé[71-73] using infrared spectroscopy (200 to 4000 cm^{-1}). NMA substituted with ^{15}N and deuterium also was studied. Two distinct sets of bands were found which were attributed to the *cis* and *trans* forms of NMA. The proportions of the two forms were found to vary with the nature of the matrix gas. In nitrogen it was concluded that there were approximately equal amounts of the two species while in argon it was suggested that the *trans* form was predominant. The spectra are consistent with a pyramidal arrangement of atoms about the NMA nitrogen atom and in some of the trapped forms the N-methyl group is probably not coplanar with the (O)CN moiety.

IV. Physical Properties and Structure of Liquid N-Methylacetamide

a) Melting Point

Various workers now generally agree that the melting point of pure NMA (purified either by zone refining or by extensive fractional freezing) is 30.55 ± 0.03 °C[13, 14, 74]. Melting points considerably lower than this have been reported by many researchers[11, 75, 76] but it is probable that in most of these cases the solvent NMA used was slightly impure. A single report[77] exists of NMA prepared with a melting point of 30.90 °C, but there appears no reason why the method of purification used by these workers should have given solvent melting more than 0.3 °C higher than the melting points of samples of solvent prepared by several other groups. In the absence of further confirmation of the value of 30.90 °C, the value of 30.55 °C is to be preferred.

It should be remembered in considering data for many of the other physical properties of NMA that the experiments were often performed using material which was slightly impure. Indeed, a considerable amount of data exists on the properties of liquid NMA at 30 °C, *i.e.*, well below the freezing point of pure NMA, although some of these data may have been obtained using supercooled rather than impure material.

b) Vapor Pressure, Enthalpy of Vaporization, and Boiling Point

The variation of the vapor pressure of NMA with temperature has been thoroughly investigated for the temperature range of 60 to 206 °C by Kortüm and co-workers[78, 79] and from 30 to 90 °C by Gopal and Rizvi[80]. Using the data of Manczinger and Kortüm[79], a value of 55.8 kJ · mol^{-1} for the enthalpy of vaporization of NMA at 100 °C can be calculated by the method described by Thompson[81]. Using approximate vapor pressure measurements between 30 and 100 °C, Dawson, Zuber and Eckstrom[82] had earlier estimated the enthalpy of vaporization to be

56 ± 8 kJ $\cdot$ mol^{-1}. The vapor pressure data indicate that the boiling point of NMA is 206.7 ± 0.1 °C at a pressure of 760 Torr.

c) Viscosity

Although many workers have reported values for the viscosity of NMA at 40 °C, or near 30 °C, only a few studies have been done which cover any sizeable range of temperature. The results of Dawson, Sears and Graves[11] (30 to 60 °C) are in substantial agreement with those of Gopal and Rizvi[83] (30 to 90 °C) and of Gopal and Rastogi[84] (35 to 55 °C). Results showing good agreement have been obtained at single temperatures within this range by several other groups[14, 85−87]. All of these results may be summarized (T in K) with an average error of 0.33% by:

$$\eta = \exp(-30.64 + 2.683(10^4/T) - 8.184(10^6/T^2) + 9.069(10^8/T^3)) \qquad (1)$$

The values of Casteel and Amis[88] are only very slightly lower (30.5, 40.0 and 50.0 °C). A value obtained by Assarsson and Eirich[89] at 30 °C agrees fairly well with the extrapolation of values obtained at 35 and 45 °C by French and Glover[76] but these results are all markedly ($\simeq 6\%$) lower than the corresponding data given by Eq. (1). These results are particularly puzzling in that viscosity values do not appear to be unusually sensitive to the purity of NMA (the viscosities noted at 40 °C in Refs.[11] and [14] agree quite well although the solvent used in the work described in Ref.[14] was undoubtedly of much higher purity).

Results of Basov, Karapetyan and Krysenko[90] for the viscosity of NMA at 25(!), 50 and 75 °C are even lower and appear to be in error.

The activation energy of viscous flow for NMA can be calculated[91] to be 19.2 ± 0.2 kJ $\cdot$ mol^{-1} at 40 °C from the constants in Eq. (1).

d) Self-Diffusion Coefficient

Williams, Ellard and Dawson[92] have used the capillary method with ^{14}C-labelled NMA to determine the self-diffusion coefficient of NMA. Values were found to range from 4.11×10^{-10} m$^2 \cdot$ sec^{-1} at 35 °C to 7.33×10^{-10} m$^2 \cdot$ sec^{-1} at 60 °C. The activation energy for the self-diffusion of NMA was calculated to be 19 kJ mol^{-1}. This is quite close to the value of 19.2 kJ $\cdot$ mol^{-1} calculated for the activation energy of viscous flow, thus suggesting the same rate controlling step for the two processes.

e) Dielectric Constant

Perhaps the most striking of the physical properties of liquid NMA is its very high static dielectric constant (initially determined by Leader and Gormley[1]), which has a value greater than 170 near its melting point. Considering the fact that the vapor phase dipole moment has a value of only 3.71 D[23], it is generally agreed that extensive ordering of the NMA molecules must occur in the liquid to account for the high dielectric constant.

Values for the dielectric constant have been determined over a wide range of temperatures by several different research groups[11, 26, 93, 94] and at one temperature by a number of others[14, 87, 95]. In general, dielectric constant data reported by the different investigators show good agreement — particularly those resulting from the extensive studies of Bass et al.[93] and of Lin and Dannhauser[26].

The temperature dependence of the dielectric constant, as determined from the values reported by many groups of workers, can be summarized (with an average percentage deviation of 0.7%) by Eq. (2):

$$\epsilon = 17.5_1 - 4.614_6(10^4/T) + 2.884_0(10^7/T^2) \qquad (T \text{ in K}) \qquad (2)$$

The values for the dielectric constant of NMA reported by Lutskii and Mikhailenko[94] are generally lower than those found by other workers at temperatures below 80 °C and markedly higher at temperatures above 100 °C. Bonner and co-workers[13, 17] have reported that freshly melted zone-refined NMA has a markedly higher dielectric constant at 40 °C than has been noted by other investigators (184 as compared to 164). The dielectric constant of such NMA is reported by Bonner et al.[13] to have decreased slowly, with time, to a final equilibrium value of 165. A value of 191 also was determined at 32 °C (compared to a value of 176 interpolated from the values of other workers). These different values suggest that NMA, in its purest obtainable form in the liquid state, tends to decompose slightly with time until some type of equilibrium is established.

The temperature dependence of the dielectric constant is very marked, with the value of the constant decreasing considerably more rapidly with increasing temperature than does the value of the dielectric constant for the isomeric N,N-dimethyl-formamide[94]. This decrease is enhanced by the breaking of hydrogen bonds as the temperature is increased, i.e., the breaking down of the chain ordering which is believed to account for the high value for the dielectric constant of NMA near room temperature.

Dielectric relaxation of NMA in the frequency range of 1 to 250 MHz has been examined from 31.4 to 84.9 °C by Bass et al.[93], and relaxation times have been determined. Itoh et al.[96] have done similar measurements at 0.356 to 2.10 GHz at 31.4 °C. They determined that all the dielectric data for NMA could be reasonably represented by the simple Debye equation

$$\epsilon' - j\epsilon'' = \epsilon_\infty + (\epsilon_0 - \epsilon_\infty)/(1 + jw\tau) \qquad (3)$$

f) Refractive Index

The refractive index of NMA has been measured by several workers over a very limited range of temperatures. Kortüm and Biedersee[78] report the refractive index of liquid NMA near the melting point as a function of temperature as follows, with t in °C from 22 to 32 °C:

$$n_D^t = 1.4325 - (4 \times 10^{-4})(t - 22) \tag{4}$$

This leads to values which agree well with those obtained at 30 °C[95] and 32 °C[23] and is not inconsistent with the value of 1.4257 obtained at 40 °C by Sears et al.[14]. The value reported by Lin and Dannhauser[26] at 31.5 °C appears to be in error.

g) Density

As with the viscosity measurements, a number of determinations exist for the density of NMA at one to three temperatures[14, 26, 76, 79, 85–90, 95] but only two studies[11, 83] exist in which the density at more than three temperatures was examined. The agreement among the results of most of the determinations is quite good and the density in $kg \cdot m^{-3}$ may be expressed as the following function of temperature expressed in °C:

$$\rho = (0.9741_3 - 0.000806_6 \, t) \, 10^3 \; kg \cdot m^{-3} \tag{5}$$

The data of Basov et al.[90] do not appear to be of precision comparable to that in the other studies noted. Densities of NMA at 60°, 80° and 100 °C, calculated from the molar volume data of Manczinger and Kortüm[79], are not consistent with the above equation.

Gopal and Rizvi[83] have used the temperature dependence of the density to determine an approximate value of 690 °K for the critical temperature of NMA.

h) Correlation Factor (g)

The availability of experimental dipole moment (Section III-a), dielectric constant, refractive index and density data for NMA has made possible the calculation of values of the correlation factor (g) using the following equation of Kirkwood[97] and Fröhlich[98]:

$$g = [(\epsilon - \epsilon_\infty)(2 \, \epsilon + \epsilon_\infty) 9 \, MkT]/[\epsilon(\epsilon_\infty + 2)^2 4 \, \pi N \rho \mu_0^2] \tag{6}$$

For non-associated compounds with random intermolecular forces, $g = 1$. If there is antiparallel alignment of molecular dipoles, $g < 1$; for parallel alignment of such dipolar molecules, $g > 1$. Several groups of workers[26, 93, 94] have noted that NMA near its freezing point has a value of g which is considerably greater than 4 and that this value decreases markedly with increasing temperature (although the value of g is still approximately equal to 3 even near the boiling point of NMA). This suggests extensive association of the amide in the pure liquid, with the dipoles in parallel alignment. The lower g values at higher temperatures can be attributed to decreased chain association of the molecules of NMA.

Lutskii and Mikhailenko[99] attempted a theoretical calculation of the correlation factor based on a quasi-crystalline model for liquids. To obtain a reasonable degree of agreement with experimental results, they found it necessary to consider that the dipole moment of an NMA molecule is increased substantially when hydrogen bonding occurs.

i) Surface Tension

Gopal and Rizvi[83] have obtained values for the surface tension of NMA at several temperatures ranging from 30 to 90 °C. Somewhat higher values have been obtained at 35 °C (36.4 versus 32.9 dynes $\cdot$ cm^{-1})[100] and 30 °C (40.1 versus 33.6 dynes $\cdot$ cm^{-1})[101] by other workers. These values are similar to those for many organic liquids. The value of the work of cohesion[102] also has been obtained at 40 °C from the surface tension data. It is markedly lower than the activation energy of viscous flow (9.9 kJ $\cdot$ mol^{-1} compared to 19.2 kJ $\cdot$ mol^{-1}), again suggesting molecular association in liquid NMA.

j) Enthalpy of Fusion and Cryoscopic Constant

A value of $(1.330 \pm 0.007) 10^5$ J $\cdot$ kg^{-1} has been determined by Kreis and Wood[103] for the enthalpy of fusion of NMA. This value leads to a calculated cryoscopic constant of 5.77 ± 0.02 K $\cdot$ mol^{-1} $\cdot$ kg. This calorimetry-based cryoscopic constant is in fair agreement with the values obtained both by Wicker[104] (5.98 ± 0.12) and by Bonner and co-workers[13, 17] (6.0 ± 0.2) from freezing point depression data, provided only fairly concentrated solutions (>0.05 M) are considered. An earlier value for the enthalpy of fusion[13] is probably in error because of solvent decomposition and Bonner and Woolsey[17] have concluded that "NMA is not suitable as a solvent in which exact measurements of colligative properties are to be made".

k) Minimum Specific Conductance

The lowest reported value for the specific conductance of NMA is 3.3×10^{-7} ohm$^{-1} \cdot$m^{-1} at 40 °C[16]. A similar value of 5×10^{-7} ohm$^{-1} \cdot$ m^{-1} also has been reported at 35 °C[105]. While the specific conductance of "absolutely pure" NMA may be lower, specific conductance values in the range of those noted above represent the lowest presently attainable.

l) Spectroscopic and Other Structural Studies

The results of spectroscopic studies of NMA as a gas, as a solid and in solution in other solvents, have been discussed previously. Infrared and Raman[25, 65, 67, 68, 106], UV[107], and NMR[108−110] spectra have been reported for pure liquid NMA. There appears to be general agreement that liquid NMA is predominantly, if not entirely, in the *trans* configuration. An attempt[111] to determine the barrier to rotation about the C—N bond using Raman spectroscopy resulted in a value of about 56 kJ · mol^{-1} but this value is probably low[57]. Comparison of the vibrational spectra of the pure liquid with those obtained for NMA as a solute in dilute solutions in other solvents suggests that hydrogen bonding in the pure liquid promotes strong chain association of NMA molecules.

Bonner and Kim[112] have reported an osmometric study which suggests that the average chain length in pure NMA is about five monomeric NMA units. Wood and DeLaney[113] have inferred an average length of fifteen monomeric units from their studies of gas solubilities in NMA. Lin and Dannhauser[26] have calculated ΔH (−19 kJ · mol^{-1}), ΔS (31 J · mol^{-1} · K^{-1}), and ΔG (−9 kJ · mol^{-1}) for hydrogen bond formation in liquid NMA (298 K) using the temperature dependence of the Kirkwood correlation factor.

Gopal and Rizvi[114] have concluded from viscosity, surface tension and vapor pressure measurements that the dispersion energy makes a larger contribution to the cohesive energy of NMA than does the hydrogen bonding energy. The sum of the dipole-dipole and dipole-induced dipole energies is quite small by comparison.

V. Electrochemistry

a) Electrolytic Conductance

An examination of the electrical conductance of certain salt solutions was the focus of the first studies using NMA as an electrolytic solvent[11, 76]. Indeed, the conductance properties of such solutions has remained one of the most frequently examined facets of the solvent properties of NMA. Much of the work has been done at 40 °C and a comprehensive summary of the Λ_0 values for salts in NMA solution at this temperature is presented in Table V-1. The Λ_0 values for a number of salts which have been studied at several temperatures, or only at one temperature which was not 40 °C, are listed in Table V-2.

The values quoted are usually those in the original papers — this despite the fact that different conductance equations may have been used to facilitate the extrapolation of the sets of data to obtain values for Λ_0. In general, it has been noted that even a simple phoreogram, i.e., a plot of Λ versus $\sqrt{C}$, extrapolated to $\sqrt{C}$ equal to zero will yield a value for Λ_0 which will not differ by more than 1% from Λ_0 values derived from more precise conductance equations, provided association is not extensive.

The value of the dielectric constant is incorporated into precise conductance treatments. However, as noted previously, there is some disagreement concerning the value of the dielectric constant of pure NMA. Most authors have used values of the dielectric constant consistent with the work of Bass et al.[93] and that of Lin and Dannhauser[26]. Alternatively, Kortüm and co-workers have used Bonner's higher value of ϵ for NMA at 35 °C[13] in conjunction with the Fuoss-Onsager equation[128] to treat data for several salts. This approach has only very slightly affected the deter-

Table V-1. Limiting equivalent conductances in NMA at 40 °C

Electrolyte[1]	$\Lambda_0 \times 10^4$ ohm$^{-1}\cdot$m$^2\cdot$eq^{-1}	Ref.
LiCl	18.1	11)
LiClO$_4$	23.35	85)
NaCl	19.6	11)
NaBr	21.0	11)
NaI	22.8	11)
NaClO$_4$	24.95	115)
NaSCN	24.14	115)
NaNO$_3$	22.65	115)
NaBrO$_3$	21.82	115)
Na picrate	20.17	115)
Na oleate	21.00	116)
Na stearate	22.00	116)
Na palmitate	27.00	116)
Na deoxycholate	27.00	116)
Na(O$_3$SPh)	18.37	115)
Na(OAc) $\cdot$ 3 H$_2$O	17.94	117)
Na(C$_6$H$_5$NHC$_6$H$_4$SO$_3$)	15.27	117)
Na benzoate	17.22	16)
Na butyrate	17.38	16)
Na cinnamate	16.15	16)
Na p-nitrophenoxide	18.27	16)
Na 2,4-dinitrophenoxide	19.53	16)
Na propionate	17.66	16)
Na 2, 4, 6 trichlorophenoxide	18.54	16)
Na phenoxide	17.5	118)

Electrolyte[1]	$\Lambda_0 \times 10^4$ ohm$^{-1}\cdot$m$^2\cdot$eq^{-1}	Ref.
Na trimethylacetate	17.02	16)
Na valerate	16.94	16)
KCl	20.1	119)
KBr	21.1	11)
KI	23.0	120)
KClO$_4$	25.22	115)
KSCN	24.47	115)
KNO$_3$	22.92	115)
KBrO$_3$	21.95	115)
K picrate	20.18	115)
K(OctdSO$_4$)	15.45	115)
KOAc	18.2	119)
RbI	23.58	121)
RbClO$_4$	25.78	85)
CsI	24.28	121)
CsClO$_4$	26.38	85)
HCl	20.68	115)
H picrate	20.87	115)
H 3-amino-benzenesulfonate	17.42	16)
H 4-amino-benzenesulfonate	17.5	16)
H dichloroacetate	19.1	122)
AgClO$_4$	25.25	85)
NH$_4$Cl	21.18	123)
NH$_4$Br	22.52	123)
NH$_4$I	24.27	123)

Table V-1 (continued)

Electrolyte[1]	$\Lambda_0 \times 10^4$ ohm^{-1}·m^2·eq^{-1}	Ref.
NH$_4$NO$_3$	24.24	115)
NH$_4$ClO$_4$	26.45	115)
Me$_4$NI	26.66	123)
	26.75	75)
Me$_4$NClO$_4$	28.77	85)
Me$_4$NBr	24.79	123)
Me$_4$NCl	23.58	123)
Et$_4$NBr	24.47	123)
Et$_4$NI	26.21	123)
	26.25	75)
Et$_4$NClO$_4$	28.52	85)
Pr$_4$NI	23.73	123)
	23.80	75)
Pr$_4$NBr	22.01	123)
Bu$_4$NI	22.34	123)
	22.50	75)
Bu$_4$NClO$_4$	24.47	85)
Pent$_4$NI	22.00	75)
Hex$_4$NI	21.80	75)
Hep$_4$NI	21.47	75)
MeNH$_3$ClO$_4$	28.58	85)
MeNH$_3$Cl	23.43	123)
MeNH$_3$Br	24.69	123)
Me$_2$NH$_2$Cl	25.22	123)
Me$_2$NH$_2$Br	26.44	123)
Me$_3$NHClO$_4$	29.57	85)
Me$_3$NHI	27.68	124)
Me$_3$NHBr	25.72	123)
Me$_3$NHCl	24.40	123)
EtNH$_3$Cl	22.38	123)
EtNH$_3$Br	23.72	123)
Et$_2$NH$_2$Cl	23.45	123)
BuNH$_3$Cl	21.33	123)
Bu$_2$NH$_2$Cl	20.89	123)
Me$_3$PhNCl	21.80	123)
Me$_3$PhNI	24.80	123)
	24.89	124)
Me$_3$PhN(O$_3$SPh)	20.46	115)
Me$_3$CeNBr	21.88	124)
OctdMe$_3$NI	21.72	115)
OctdMe$_3$NNO$_3$	21.60	115)
OctdMe$_3$N(OctdSO$_4$)	14.26	115)
MgCl$_2$ · 6 H$_2$O	21.33	125)
Mg(NO$_3$)$_2$ · 5 H$_2$O	24.20	125)
Ca(NO$_3$)$_2$ · 4 H$_2$O	24.69	125)
Ca(ClO$_4$)$_2$ · 4 H$_2$O	27.00	125)
CaBr$_2$ · 4 H$_2$O	23.13	125)

Table V-1 (continued)

Electrolyte[1]	$\Lambda_0 \times 10^4$ $ohm^{-1} \cdot m^2 \cdot eq^{-1}$	Ref.	Electrolyte[1]	$\Lambda_0 \times 10^4$ $ohm^{-1} \cdot m^2 \cdot eq^{-1}$	Ref.
$SrCl_2 \cdot 6\,H_2O$	21.93	125)	$BaI_2 \cdot 2\,H_2O$	24.71	125)
$SrBr_2 \cdot H_2O$	23.13	125)	BaI_2	24.66	125)
$SrBr_2 \cdot 6\,H_2O$	23.13	125)	$Ba(NO_3)_2$	24.60	125)
$Sr(NO_3)_2$	24.92	125)	$Ba(ClO_4)_2$	27.07	125)
$Sr(ClO_4)_2 \cdot 3\,H_2O$	27.26	125)	$Ba(ClO_4)_2 \cdot 2\,H_2O$	27.07	125)
$Sr(ClO_4)_2$	27.30	125)	$Ba(C_6H_5NHC_6H_4SO_3)_2$	17.20	117)
$BaCl_2 \cdot 2\,H_2O$	21.64	125)	$Na_2Fe(CN)_5NO \cdot 2\,H_2O$	23.94	117)
$BaCl_2$	21.67	125)	$Na_3Fe(CN)_6 \cdot 3\,H_2O$	25.08	117)
$BaBr_2 \cdot 2\,H_2O$	22.92	125)	$K_3Fe(CN)_6$	25.40	117)
$BaBr_2$	22.93	125)	$K_3Fe(C_2O_4)_3 \cdot 3\,H_2O$	24.05	117)

[1] Me = methyl, Et = ethyl, Pr = n-propyl, Bu = n-butyl, Pent = n-pentyl, Hex = n-hexyl, Hept = n-heptyl, Octd = octadecyl, Ph = phenyl, Ce = cetyl, and OAc = acetate.

Table V-2. Limiting equivalent conductances of electrolytes in NMA at several temperatures $\Lambda_0 \times 10^4$ (ohm$^{-1} \cdot$ m$^2 \cdot$ eq^{-1})

	30 °C	35 °C	40 °C	45 °C	50 °C	55 °C	60 °C	Ref.
LiCl	14.1		18.1		22.4		27.4	11)
LiClO$_4$		20.76	23.35		24.5			85, 105)
NaCl	15.5	17.77	19.6	22.37			29.7	11, 76)
NaBr		18.91	21.0	23.77				11, 76)
NaI	17.8	20.59	22.8	25.83	28.0		34.0	11, 76)
Na oleate		18.75	21.00	23.25	25.37			116)
Na stearate		20.25	22.00	23.65	25.12			116)
Na palmitate		25.27	27.00	28.50	30.00			116)
Na deoxycholate		24.50	27.00	28.25	30.00			116)
KCl	15.8	17.88	20.1	22.57				11, 76)
	16.22 (31 °C)							126)
KBr	16.6	18.84	21.1	23.92	26.2		32.1	11, 76, 105)
		19.00						
KI	18.1	20.70	23.0	26.04	28.5			11, 76, 120)
			23.1					
RbI		21.38	23.58	26.46	28.74	31.21		121)
CsBr		20.05		25.23				76)
CsI		22.01	24.28	27.29	29.51	32.21[1]		121)
HCl		18.38	20.68					87, 115)
HBr		19.56						105)
HI		21.08						105)
Me$_4$NI		24.70	26.75	29.59	32.10	35.00		75, 123)
Et$_4$NBr		22.05	26.66	27.71				76, 123)
Et$_4$N picrate		21.26	24.47	26.75				76)
Et$_4$NI		24.10	26.25	29.15	31.45	34.35		75, 123)
			26.21					
Pr$_4$NI		21.65	23.80	26.60	28.85	31.50		75, 123)
			23.73					

Table V-2 (continued)

	30 °C	35 °C	40 °C	45 °C	50 °C	55 °C	60 °C	Ref.
Bu$_4$NI		20.53	22.50	25.20	27.35	29.80		75, 123)
			22.34					
Pent$_4$NI		20.10	22.00	24.60	26.61	29.02		75)
Hex$_4$NI		19.90	21.80	24.41	26.34	28.72		75)
Hept$_4$NI		19.62	21.47	24.00	25.95	28.00		75)
Me$_3$NHI		25.15	27.68	31.11	34.26	37.00		124)
Me$_3$PhNI		22.40	24.89	27.95	30.50	33.05		124)
Me$_3$CeNBr		19.89	21.88	24.41	26.64	28.96		124)
H chloroacetate	5.26							127)
H dichloroacetate	13.55		19.1		20.92			122, 127)
H trifluoroacetate	16.04				24.56			127)

[1]) Value given in the paper is 31.21 but all calculations in the paper appear to have been done using 32.21.

mined values of Λ_0 but has significantly influenced the theoretical interpretation of the phoreograms for a given electrolytic type.

For most of the salts the slope of the phoreogram is steeper (*i.e.*, more negative or catabatic) than would be calculated by the Onsager limiting law. Most researchers have rejected the possibility of ionic association as a cause of this deviation – principally because the Bjerrum contact distance in NMA, for the temperature range of 30 to 60 °C, is less than 0.2 nm and thus less than the sum of the crystallographic radii of most possible cation-anion contact pairs.

Recent refinements of conductance theory by Fuoss[129, 130] plus the indication that certain salts can be moderately associated in other high dielectric constant solvents[131] may suggest a need to re-examine the possibility of ion-pair formation for some salts in NMA.

More usually the catabatic phoreograms have been considered a result of changes in the solution viscosity with increasing salt concentration. Thus Dawson *et al.*[11] obtained linear plots up to concentrations of 0.4 M by plotting $\Lambda\sqrt{\eta}$ versus $\sqrt{C}$ and Kortüm and Hebestreit[105] have defined $\Lambda\eta \equiv \Lambda(1 + BC)$ and obtained linear plots of $\Lambda\eta$ versus $\sqrt{C}$ to concentrations up to 0.1 M. Whatever the merits of these treatments, there is no denying that moderate salt concentrations make large changes in the viscosity[11] and apparent dielectric constant[132] relative to the values for the pure solvent.

As water is a possible impurity in solvent NMA, it is somewhat reassuring that values for Λ_0 for some of the hydrated and anhydrous alkaline earth salts agree so well. This also suggests that the ions in question are probably solvated moderately well in dilute solution by NMA molecules in competition with water molecules. Small quantities of acetic acid – also a common impurity in NMA – also have been shown to have little effect on conductance measurements for most simple 1–1 and 2–1 electrolytes; however, this is not the case for solutions of salts of many transition and inner-transition metals[119].

The Λ_0 value for monochloroacetic acid reported by Fialkov and Suprunenko[127] appears to be extremely low and is probably in error.

Rastogi has examined the conductance of 0.1 M solutions of several electrolytes near the melting point of NMA[133]. The specific conductances were found to decrease smoothly for the liquid solutions as the temperature was lowered from 35 to 27 °C. Activation energies of electrical conduction were calculated and were found to be similar to those for viscous flow. Specific conductances were also determined for similar 0.1 M solutions over a temperature range of 0 to 50 °C[134]. Below about 25 °C solid solutions were formed. On solidification of the solutions the specific conductances, as might be expected, were found to drop markedly. The specific conductances for the solutions of a series of potassium halides were found to follow the same order in solid NMA as in liquid NMA (KI > KBr > KCl). However, a series of solid solutions of iodide salts were found to have specific conductances following the order $NaI > KI > CsI > Et_4NI$, which is exactly the reverse order to that found for the iodide salts in liquid NMA.

b) Ionic Transport

Ideally, to determine limiting ionic conductances in a solvent, precise transport number data are required for an unassociated salt for which the value of Λ_0 is known accurately. Several attempts now have been made to determine transport values, by the Hittorf method, for ions in NMA using a variety of salts[85, 86, 135]. A summary of values obtained for $\lambda^+_{0_{K^+}}$ in NMA at 40 °C is provided in Table V-3.

Table V-3. Limiting ionic equivalent conductance of K^+ in NMA from transport measurements

Salt	$t^0_{K^+}$	$10^4 \lambda^+_{0_{K^+}}$ ($ohm^{-1} \cdot m^2 \cdot eq^{-1}$)	Ref.
KBr	0.3900	8.23	135)
KCl	0.4292	8.62	86)
KSCN	0.3475	8.50	85)

The values show a disappointing lack of agreement. Part of the problem is the result of some inaccurate Λ_0 values. Thus $\Lambda_{0\,NaSCN} - \Lambda_{0\,NaCl} = 4.54 \times 10^{-4}$ $ohm^{-1} \cdot m^2 \cdot eq^{-1}$ but $\Lambda_{0\,KSCN} - \Lambda_{0\,KCl} = 4.37 \times 10^{-4}$ $ohm^{-1} \cdot m^2 \cdot eq^{-1}$, a difference in the results of almost 4% which may or may not be correlated with a single erroneous Λ_0 value. Also $\Delta\Lambda_0$ between the values for chloride and bromide salts of a common cation generally is found to be $(1.2-1.3) \times 10^{-4}$ $ohm^{-1} \cdot m^2 \cdot eq^{-1}$ in NMA at 40 °C but $\Delta\Lambda_0 = 1.0 \times 10^{-4}$ $ohm^{-1} \cdot m^2 \cdot eq^{-1}$ for KCl and KBr suggesting a rather large error in the Λ_0 value of one of these two salts which has a direct bearing on the λ^+_0 values in Table V-3.

Further, Paul and co-workers[85] have presented transport data for $AgClO_4$ in NMA at 40 °C. They claim close agreement of λ^0_0 values calculated using $t^0_{Ag^+}$ from these experiments with values calculated using $t^0_{K^+}$ from the KSCN experiments[85]. Examination of the data presented in Ref.[85] suggests that using the Λ_0 values for $KClO_4$ and $AgClO_4$ listed in Table V-l, $\lambda^+_{0_{K^+}}$ and $\lambda^-_{0_{ClO_4}}$ have values of 8.58×10^{-4} and 16.67×10^{-4} $ohm^{-1} \cdot m^2 \cdot eq^{-1}$, respectively, in NMA at 40 °C. These values differ by about 1% from values derived from the KSCN results and by less than 1% from the results of Johari and Tewari[86]. (Indeed, a value of $\Lambda_{0_{KCl}} = 19.9_5 \times 10^{-4}$ $ohm^{-1} \cdot m^2 \cdot eq^{-1}$ — a value which is reasonable when the limiting conductance values for many salts are compared — would give $\lambda^+_{0_{K^+}} = 8.56 \times 10^{-4}$ $ohm^{-1} \cdot m^2 \cdot eq^{-1}$, in fairly close agreement with the work of Paul et al.[85].

Considering the inherent errors in Λ_0 and the different conductance equations used in obtaining the Λ_0 values, the $\lambda^+_{0_{K^+}}$ value determined by Gopal and Bhatnagar[135] still appears to differ significantly from the other values noted. There may be several reasons for this. First of all, the purity of the solvent used by Gopal and Bhatnagar (mp. = 29 °C) is suspect. Secondly, the transport data (on KBr solutions) were obtained using relatively concentrated solutions (up to 0.30 M with no solutions of a concentration less than 0.10 M), and the Longsworth function used for extrapolation to obtain t_+^0 may not be valid at these higher concentrations[136]. Also the method used by these workers to analyze for potassium has been criticized by Notley and Spiro[137].

Probably a value of $\lambda_{0_{K^+}}^+ = 8.54 \times 10^{-4}$ ohm$^{-1} \cdot$ m$^2 \cdot$ eq^{-1} in NMA at 40 °C is correct within 1.5% and this value has been used in this review to prepare a table of limiting ionic conductances in NMA at 40 °C (Tabel V-4). It should be noted in passing that the approximation of λ_0^+ (octadecyltrimethylammonium ion) = λ_0^- (octadecylsulfate ion)[115] gives $\lambda_{0_{K^+}}^+ = 8.32 \times 10^{-4}$ ohm$^{-1} \cdot$ m$^2 \cdot$ eq^{-1} at 40 °C, which agrees within 3% with the value from the transport experiments.

The only experimental work on transport numbers at temperatures other than 40 °C has been done by Gopal and Bhatnagar[135] and this is subject to the same problems noted above. However, these problems would not be expected to affect the general trends noted, namely, that $t_{K^+}^0$ increases with increasing temperature and that $t_{K^+}^0$ decreases markedly with increasing KBr concentration. Krumgal'z' estimated values for $\lambda_0^{\pm}$ in NMA at 40 °C[138] are inconsistent with all of the experimental studies and there is no reason to assume that values of $\lambda_0^{\pm}$ estimated for other temperatures by this investigator are not similarly in error by 20 to 30%.

c) Limiting Ionic Conductances

Several interesting trends are apparent on examining the data in Table V-4. For simple, singly-charged cations the limiting ionic conductances follow the order:

$$\text{Li}^+ < \text{Na}^+ < \text{K}^+ \simeq \text{Ag}^+ < \text{Rb}^+ \simeq \text{H}^+ < \text{Cs}^+.$$

This order suggests that the smaller alkali metal cations interact quite strongly with the NMA solvent so that their effective Stokes' radii are larger than those of Rb$^+$ or Cs$^+$. Similarly the values of λ_0^- for the halide ions follow the order Cl$^- <$ Br$^- <$ I$^-$. It is also clear that there is no abnormally large limiting ionic conductance for the hydrogen ion in NMA solution.

Interesting orders are found for λ_0^+ for the partially substituted alkylammonium ions where

$$\text{NH}_4^+ < \text{MeH}_3\text{N}^+ < \text{Me}_2\text{H}_2\text{N}^+ > \text{Me}_3\text{HN}^+ > \text{Me}_4\text{N}^+$$
$$\text{NH}_4^+ < \text{EtH}_3\text{N}^+ < \text{Et}_2\text{H}_2\text{N}^+ \qquad \simeq > \text{Et}_4\text{N}^+$$
$$\text{NH}_4^+ \lessapprox \text{BuH}_3\text{N}^+ > \text{Bu}_2\text{H}_2\text{N}^+ > \text{Bu}_4\text{N}^+$$

The order is the result of at least two competing factors. Where small alkyl groups are part of the partially substituted species, the ions can participate readily in hydrogen bonding with the solvent, thus decreasing λ_0^+. As more alkyl groups become involved this potential for interaction with the solvent is decreased leading to an apparent increase in λ_0^+. However, as more alkyl groups are added the effective size of the ions increase decreasing λ_0^+. Where larger alkyl groups are involved this second factor appears to predominate.

Table V-4. Limiting ionic equivalent conductances in NMA at 40 °C[1])

Cation	$10^4 \lambda_0^+$ $(ohm^{-1} \cdot m^2 \cdot eq^{-1})$	Anion	$10^4 \lambda_0^-$ $(ohm^{-1} \cdot m^2 \cdot eq^{-1})$
Li^+	6.69	Cl^-	11.41
Na^+	8.30	Br^-	12.66
K^+	8.54	I^-	14.48
Rb^+	9.11	ClO_4^-	16.66
Cs^+	9.76	NO_3^-	14.36
H^+	9.20	SCN^-	15.88
Ag^+	8.58	BrO_3^-	13.47
NH_4^+	9.82	Picrate	11.75
Me_4N^+	12.17	Octadecylsulfate	6.93
Et_4N^+	11.79	Acetate	9.65
Pr_4N^+	9.29	$PhSO_3^-$	10.07
Bu_4N^+	7.90	$C_6H_5NHC_6H_4SO_3^-$	6.96
$Pent_4N^+$	7.52	Oleate	12.70
Hex_4N^+	7.32	Stearate	13.70
$Hept_4N^+$	6.99	Palmitate	18.70
MeH_3N^+	11.99	Deoxycholate	18.70
$Me_2H_2N^+$	13.80	Benzoate	8.92
Me_3HN^+	13.04	Cinnamate	7.85
EtH_3N^+	11.02	p-Nitrophenoxide	9.97
$Et_2H_2N^+$	12.04	2,4-Dinitrophenoxide	11.23
BuH_3N^+	9.92	Propionate	9.36
$Bu_2H_2N^+$	9.48	Butyrate	9.08
Me_3PhN^+	10.38	Valerate	8.64
Me_3CeN^+	9.22	Trimethylacetate	8.72
Me_3OctdN^+	7.27	2,4,6-Trichlorophenoxide	10.24
Mg^{2+}	9.88	3-Aminobenzenesulfonate	8.22
Ca^{2+}	10.38	4-Aminobenzenesulfonate	8.3
Sr^{2+}	10.55	Dichloroacetate	9.9
Ba^{2+}	10.27	Phenoxide	9.2
		$Fe(CN)_5NO^{2-}$	15.64
		$Fe(CN)_6^{3-}$	16.82
		$Fe(C_2O_4)_3^{3-}$	15.51

[1]) The values for the limiting ionic conductances in NMA at 40 °C in this table are self-consistent and are based on $\lambda_{0K^+}^+$ equal to 8.54×10^{-4} $ohm^{-1} \cdot m^2 \cdot eq^{-1}$. The values have been calculated by considering all available conductance work for salts of each ion at 40 °C. No attempt has been made, however, to weight the results according to the precision of the experimental conductance work. All limiting conductance data determined from conductance data for salts with association constants greater than 10^3 also were not included. For ions where five or more salts have been studied, the $\lambda_0^\pm$ values probably are accurate to better than 1% exclusive of any errors in $\lambda_{0K^+}^+$.

As might be expected, the λ_0^+ values for the tetra-n-alkylammonium ions decrease fairly regularly as the number of carbon atoms in the alkyl chain is increased. Paul and co-workers have used $\lambda_0^\pm$ values to estimate apparent solvation numbers for several singly-charged ions[85]).

The λ_0^+ values for the doubly-charged alkaline earth ions vary only slightly on going from Mg^{2+} to Ba^{2+} – the order being $Sr^{2+} > Ca^{2+} > Ba^{2+} > Mg^{2+}$. While this order is not identical to that found for the ions in aqueous solution[139], the differences from ion to ion, on a percentage basis, are small – particularly for Ca^{2+}, Sr^{2+}, and Ba^{2+}.

Finally, the large values of λ_0^- for the palmitate and deoxycholate ions should be noted. The reason for these large values of λ_0^- is not at all clear and more will be said about this later (Section VII-g).

d) Behavior of Electrodes in NMA

The behavior of reference electrodes in amide solvents was reviewed several years ago by Reid and Vincent[6] who recommended the use of either a silver-silver chloride electrode or a cadmium-cadmium chloride electrode. Of these two, the former has been used in NMA by several workers[140-142].

The hydrogen electrode also has been found to perform satisfactorily in amide solvents and has been used as a reference electrode for polarographic and voltametric studies by Sibille et al.[143] Gosselet and co-workers[144] have used the electrode Ag | 0.1 M AgClO$_4$, 0.1 M LiClO$_4$(NMA) as a reference electrode where LiClO$_4$ was employed as an electrolyte added to all solutions being measured. Both Payne[145] and Knecht[140] have used aqueous calomel reference electrodes for electrochemical studies although Knecht reported a slight drift in the potential of this electrode in basic NMA solutions.

The hydrogen electrode has been shown to be stable and to change potential in a Nernstian fashion with changing hydrogen ion activity in NMA[140, 144]. Glass electrodes have also shown essentially Nernstian behavior in responding to changes in hydrogen ion activity in acidic NMA solutions but potential readings have been reported to drift in basic solutions[140]. Antimony electrodes were found to give erratic and irreproducible potentials in NMA solutions[140]. Various metal wire electrodes and the dropping mercury electrode have been used as indicator electrodes in a study of electrochemical reduction of several species in NMA[143]. Studies also have been done using a rotating platinum electrode[141, 146].

e) Electromotive Force Measurements

Dawson and co-workers[82, 142] have determined EMF values for the cell

Pt, H$_2$; HCl(m); AgCl–Ag

in NMA over the temperature range of 35 to 70 °C. These values were used to determine activity coefficients for HCl in NMA over a concentration range of 0.002 to 0.1 M. The activities are higher than those for HCl in water (but $\gamma_\pm < 1$) and approach the Debye-Hückel limiting law at low concentrations. The temperature-dependent EMF data were used to calculate the relative partial molal enthalpies and heat capacities of hydrogen chloride in NMA as well as $\Delta G°$, $\Delta H°$ and $\Delta S°$ for the cell reaction. Bonner and Woolsey[17] have expressed some doubts concerning these EMF data because the HCl would tend to accelerate the decomposition of the NMA solvent.

f) Polarography

A limited number of polarographic investigations in solvent NMA have been reported. Sellers and Leonard[147] used a dropping mercury electrode to determine half-wave potentials of a number of organic and inorganic compounds using the mercury pool as the reference electrode and using 0.12 M Et_4NBr as the supporting electrolyte. The Cd^{2+} and Pb^{2+} ions were found to undergo two-electron reductions which were essentially reversible, while well-defined polarograms could not be obtained for solutions of $CrCl_3$, $CoCl_2$ and $NiCl_2$. Irreversible reductions were indicated for p-nitro-aniline, p-nitroacetanilide, azobenzene, benzil, and 4,4'-bis(acetamido)azoxybenzene. Knecht and Kolthoff[8] determined "depolarization potentials" for a number of simple inorganic cations and anions in NMA. The anodic potentials were determined using a supporting electrolyte, 0.1 M in the relevant anion[140]. The cationic depolarization potentials were determined with 0.1 M Et_4NClO_4 supporting electrolyte. The measurements were made using an aqueous saturated calomel electrode as a reference electrode. Under these conditions, the polarographic range of well-purified NMA is +0.3 to −2.7 volts. The depolarization potentials were found to follow essentially the same order and to be only slightly more negative than the potentials determined similarly for the same ions in water. The depolarization potential values for halide ions are more negative than those for perchlorate, suggesting the formation of halide ion complexes with mercury(I) or mercury(II) ions in NMA. This is confirmed by the work of Agurto-Cid and Machtinger[148] who have studied, polarographically, complex formation between Hg^{2+} and Cl^-, Br^-, I^-, SCN^-, and $S_2O_3^{2-}$ ions. Pucci et al.[141] have also investigated the Hg^{2+}–SCN^- equilibria. Knecht and Kolthoff[8] further found that the reduction of thallium(I) is reversible in NMA while copper(II) undergoes an apparently irreversible two-electron reduction.

Several groups have reported on the polarographic behavior of oxygen dissolved in NMA and there is a marked lack of agreement. Oxygen has been reported to give one reduction wave[147] ($E_{1/2}$ = −0.23 volt versus the mercury pool as reference electrode) or two waves ($E_{1/2}$ = −0.54 volt, − 1.6 volts versus an aqueous saturated calomel electrode[8]; $E_{1/2}$ = −0.14 volt, −1.7 volts versus a normal hydrogen electrode[143]). Sellers and Leonard[149] and Knecht and Kolthoff[8] agree that the wave(s) are irreversible. Sibille et al.[143] contend that the first wave ($E_{1/2}$ = −0.14 V) is reversible while that at −1.7 V is irreversible. This group of investigators also has demonstrated that the first one-electron reduction depends only slightly on pH ($E_{1/2}$ changes by about −10 mV per pH unit) and have concluded that the observations are consistent with the process: $O_2 + e^- \rightarrow O_2^-$. There is general agreement, however, that oxygen dissolved in NMA can be easily removed by passing nitrogen gas through the solvent.

It appears[8] that the polarographic behavior of ions in NMA is fairly similar to that in water and that, for purposes of analytical polarography, NMA offers few advantages over other solvents.

g) Other Studies of Reactions at Electrodes in NMA Solutions

Sibille et al.[143] have studied the reduction of oxygen at a platinized platinum wire electrode. The reduction was found to be slow. The half-wave potential showed a linear dependence on pH ($E_{1/2}$ changed by -60 mV per pH unit) and had a value of $+0.76$ V at pH = 0 (perchloric acid) versus a normal hydrogen reference electrode (although Fig. 3 of Ref.[143] suggests that this value may be for pH = 1). The wave heights were found to be proportional to the oxygen concentration and were pH independent. The reaction, $O_2 + 2 H^+ + 2 e^- \rightarrow H_2O_2$, was proposed to explain these observations.

Oxygen reduction was also examined at platinum and gold amalgam wire electrodes. The reduction wave, in both cases, was found to be slow over a large range of pH but fast near pH = 7.4. At this pH, the reduction was similar to the fast monoelectronic reduction previously described as occurring at the dropping mercury electrode; and O_2^- was again the proposed product of the reduction.

The reduction of benzoquinone in NMA at a platinum wire electrode has been studied[143]. In unbuffered solutions the reduction gave two fast, one-electron waves of almost equal height ($E_{1/2}$ = -0.04 V and -0.38 V versus a normal hydrogen reference electrode). It was proposed that these represented the formation of the radical anion and of the hydroquinone dianion, respectively. In buffered solutions an irreversible two-electron reduction (probably to hydroquinone) was found. It thus appears that certain radical anions may be much more stable in NMA than in water.

Pucci et al.[141] have done potentiometric studies of complex formation between SCN^- ion and Ag^+ and Hg^{2+} in NMA solutions. Thompson[149] has studied complex formation between tin(II) and Br^- ions in NMA potentiometrically.

h) Electrical Double Layer

Payne[145] has reported measurements of the capacity of the electrical double layer at a growing mercury drop in NMA (with a normal calomel electrode as reference electrode). Two capacity humps on opposite sides of the potential of zero charge (pzc) were found for NMA solutions of KPF_6, $LiBF_4$, $LiClO_4$ and Et_4NClO_4. The anodic hump was absent in solutions of Pr_4NBr, NH_4Cl, KNO_3 and CsI. Specific adsorption for anions from NMA onto the mercury in the order $I^- > Br^- > Cl^- > PF_6^- > (ClO_4^-, NO_3^-, BF_4^-)$ is indicated by the degree of negative shift of the pzc for salts containing these ions. The pzc of KPF_6 solutions becomes more negative with increasing salt concentration, implying adsorption of the PF_6^- anion, but electrocapillary measurements failed to provide confirmatory evidence. Specific adsorption of Cs^+, NH_4^+ and K^+ is also suggested by the capacity curves.

i) Electrolysis and Electrodeposition

Couch[150] has reported isolation of 3,7-diaza-5-oxa-2,8-nonanedione from the products of the electrolysis of NMA at bright platinum electrodes ($0.1 \ A \cdot dm^{-2}$ for $2 \ F \cdot mol^{-1}$). This suggests that the solvent decomposition by electrolysis is quite complex.

Electrolysis of NMA solutions of KNO_3, HNO_3, KNO_2, NaF, Li_2SO_4 and NH_4F have been found to give anodic deposition of oxide films on silicon single crystals[151—154]. Electrolysis of NMA solutions of KCl and NaCl did not produce such films. Examination of the anode electrolysis products showed the presence of water and of an amide (not NMA) and traces of other (unidentified) compounds. Harrick[155] has noted that the SiO_2 films appear to contain small amounts of NMA. The effects, on the deposition of the oxide layer, of small amounts of water in the NMA solvent were also determined[153].

Similar treatment of crystals of tantalum and germanium[151] and of GaAs[156] and InSb[157] as anodes in NMA apparently also leads to oxide film formation.

Panzer[158] has been granted a patent which describes the use of salt solutions in NMA as battery cell electrolytes.

VI. Solution Thermodynamics

a) Solubilities of Non-Electrolytes

Dawson et al.[159] found that many organic materials such as dioxane, cyclohexane, pyridine, toluene, nitrobenzene, chloroform, ethanol, etc., are soluble to more than 30% by weight in NMA at 40 °C. Even n-pentane is soluble to the extent of 5% by weight. Cellulose acetate is soluble to at least 0.5% by weight[150] while polyacrylonitrile is essentially insoluble in NMA[161].

Wood and DeLaney[113] have studied the solubilities of helium, nitrogen, argon and ethane over the temperature range of 35 to 70 °C. The gases are much more soluble in NMA than in water, but are not quite as soluble as in non-polar solvents. From the solubilities at 35 °C, values of ΔG, ΔH, ΔS and ΔC_p were deduced. A plot of ΔH_v ($= -\Delta H_{soln}$) versus ΔS_v ($= -\Delta S_{soln}$) gives a line which is quite similar in slope to, and lies only slightly above, the line for non-polar solvents. The results are markedly different from those for water. Also the changes in heat capacity upon dissolving gases in NMA are very small — unlike the much larger values obtained from studies of aqueous solutions.

These observations suggest that, when non-polar solutes are dissolved in NMA, the chains of NMA molecules are not appreciably disrupted and the solute fits in between the chains. The solute molecules probably tend to remain in contact with the less polar portions of the NMA chain.

Sibille et al.[143] have reported a value of 9×10^{-3} M for the solubility of oxygen in NMA. Bonner and co-workers[162] have measured the solubility of anthraquinone in NMA over the temperature range of 32 to 62 °C. The solubility data then were used to calculate the enthalpy and entropy of solution. Consideration of the entropy in conjunction with the entropy of freezing of pure NMA and with the densities of the NMA-anthraquinone solutions led to an estimate of about 2.5 as a lower limit of the solvation number of anthraquinone in NMA.

b) Solubilities of Electrolytes

Two extensive quantitative studies of the solubilities of salts in NMA have been reported. Dawson et al.[159] report accurate solubilities for fourteen salts and also approximate solubilities for twenty-four others (all at 40 °C). Chandra and Gopal[163] have determined solubilities for twenty-two different salts at each of five different temperatures (from 32 to 50 °C). The salts dissolve slowly, as pointed out by Gopal and Bhatnagar[135], and this has led some workers to underestimate the solubilities.

The measurements by the two groups noted above are in reasonably good agreement except for a few of the salts for which Dawson et al. reported only approximate values[a]. One notable exception is the solubility of NH_4Br at 40 °C where a discrepancy of approximately 20% exists. Differences in the quality of the NMA used and in the length of time allowed for the solution to come to equilibrium may be the source of different results for a single salt at a given temperature. The results of Chandra and Gopal[163] suggest a regular change of solubility with temperature for all salts which they report and, hence, these values may be considered more reliable than those of Dawson et al.[159]. Nevertheless, the times allowed by Chandra and Gopal[163] for equilibrium of the salt solutions were quite short (only a few hours). The possibility that this may have led to systematic errors in the solubility values for some of the salts cannot be disregarded.

Generally the solubilities of the salts are somewhat lower than the corresponding solubilities in water and somewhat higher than those in most dipolar aprotic solvents. The alkali metal chlorides are considerably less soluble than the perchlorates — the reverse of the trend in water. In general, salts of multiply-charged anions (particularly sulfates and carbonates) tend to be fairly insoluble.

Unfortunately, quantitative solubility data were obtained primarily for salts which are fairly soluble in NMA. Thus, values for the solubility products (and hence for the free energies of solution at infinite dilution) cannot be calculated from the solubility data with any degree of confidence as activity data are unavailable at the relevant concentrations. Nor can the temperature-dependent data of Chandra and Gopal be used to derive meaningful values for $\Delta H°$ (soln) for similar reasons. Pucci et al.[141] have determined the solubility product of AgSCN in NMA to be 2×10^{-14} at 40 °C.

---

a) The rather large disagreement[5, 159, 163] in the value for the solubility of $NaNO_3$ in NMA at 40 °C has been traced to typographical errors in Refs.[5] and [159]. The experimental value for the solubility should have been reported as 1.65 moles per liter.

c) Ionic Free Energies of Transfer from Water to NMA

Pucci et al.[141] have done an interesting study which led to approximate values for the free energies of transfer from water to NMA (0.1 M Et_4NClO_4) for the ions listed in Table VI-1. The values were calculated using the extrathermodynamic assumption of Strehlow[164] that the potential of the ferrocene/ferrocinium couple remains unchanged on transfer from one solvent to another. This approximation has been criticized for several reasons[165–167] but still leads to values which are within 10 to 15 $kJ \cdot mol^{-1}$ of the values obtained by other reasonable extrathermodynamic assumptions[166]. It is apparent that the cations studied are more strongly solvated in NMA than in water while the thiocyanate ion is less strongly solvated in NMA.

Table VI-1. Free energies of transfer from water to NMA (40 °C, 0.1 M Et_4NClO_4) (Assumption: $\log{}^0\gamma^s_{ferrocene} = \log{}^0\gamma^s_{ferrocinium\ ion}$)

Ion	ΔG_{tr} (kJ $\cdot$ mol^{-1})
Ag^+	−27
H^+	−33
Hg^{2+}	−55
SCN^-	+42

d) Enthalpies of Solution

Enthalpies of solution have been determined by Somsen and co-workers[77, 168–170] for a number of salts in NMA. The reported experimental values are given in Table VI-2. Except for KF, the enthalpies of solution of the noted alkali metal halides are more exothermic than the enthalpies of solution in water. For the tetraalkylammo-

Table VI-2. Experimental enthalpies of solution (kJ $\cdot$ mol^{-1}) for salts in NMA

Salt	35 °C	45 °C	55 °C	Refs.
LiI	−89.62			77)
NaI	−30.24			77)
KF	−8.80			77)
KCl	4.96			77)
KBr	0.789			77)
KI	−9.79			77)
RbI	−2.42			77)
CsI	7.54			77)
Me_4NBr	29.04	29.60	29.84	169)
Et_4NBr	25.49			168)
Pr_4NBr	23.74			168)
Bu_4NBr	24.42	24.83	25.35	168, 169)

nium bromides the heats of solution in NMA are less exothermic than in water. This difference arises from the so-called "solvophobic" interaction of the tetraalkylammonium ions with water[168]. De Visser and Somsen suggest that in contrast to the behavior of these ions in water, in methanol, and even in formamide, the tetraalkylammonium ions in NMA merely appear to interact with the solvent in the same manner as large alkali metal ions. Tomus[171] has combined literature lattice energy data and (aqueous) heat of solution data with the enthalpies of solution in NMA as determined by Weeda and Somsen[77] and has calculated heats of solution in NMA for all the common alkali metal halides.

The enthalpy of solution values determined at different temperatures, by analogy with the results for N-methylformamide, suggest[170] a slight degree of breaking of the NMA solvent structure by the two tetraalkylammonium bromide solutes studied.

e) Cryoscopic Measurements

Cryoscopic measurements in NMA have been the subject of investigations by Bonner and co-workers[13, 162, 172, 173] and by Wood and co-workers[74, 174]. Bonner and Woolsey[17] have concluded that "it appears that NMA is not suitable as a solvent in which exact measurements of colligative properties are to be made". This conclusion was based primarily on the slight autodecomposition of the solvent which has been discussed previously in this review. Certainly the earlier work by Bonner et al.[13, 162] led to values of osmotic coefficients which showed rather unusual concentration dependence. Most of the studies of these researchers dealt with freezing point depressions produced in NMA by a large variety of nonelectrolytes although values for a number of electrolytes also were determined. Wood, Wicker and Kreis[74, 174] have determined osmotic coefficients for a large number of electrolytes in NMA using cryoscopy. For NaI and KI they have compared their results with those of Bonner, Kim and Torres[173] over the concentration range of 0.005 M to 0.1 M. The results from the two groups of workers are consistent (within 2%) provided the same value of the cryoscopic constant (as calculated from the heat of fusion of NMA determined by Kreis and Wood[103]) is used. In general, the concentrations of salts (up to 0.8 M) used by Wood and co-workers were greater than the solute concentrations employed by Bonner and co-workers. This minimized the effect of any slight lowering of the solvent melting point caused by solvent autodecomposition. However, the use of higher concentrations also leads to larger errors if a value of the cryoscopic constant for NMA is obtained by extrapolation.

Despite the noted problems, the work of Bonner, Bunzl and Woolsey[162] suggests that the extent of disruption of the structure of solvent NMA by added nonelectrolytes is markedly dependent on the solute. Certain solutes which can act as electron acceptors (e.g., anthraquinone[162] and 4,4'-dinitrodibenzyl[172]) are found to show deviations from Raoult's law which are considerably less positive than deviations found for most non-electrolyte solutions in NMA.

Wood et al.[74] have found that the osmotic coefficients for alkali metal halides and nitrates in NMA are much higher than those for the same salts in water. This is attributed to the higher dielectric constant of NMA. Nevertheless the order of the osmotic coefficients for these salts is very similar to the order found for aqueous solution[175]. Generally the anion order is $I^- > Br^- > Cl^- > NO_3^-$ and the cation order is $Li^+ > Na^+ > K^+ > Cs^+$. Thus it is likely that the same factors are responsible for the values of the activity coefficients for these solutes both in water and in NMA.

Kreis and Wood[174] studied the osmotic coefficients of some tetraalkylammonium halides in NMA. The coefficients vary with anion in the order $Cl^- > Br^- > I^-$, the same order as found in water[176] for such salts but the reverse order to that previously noted for the alkali halides. The order found for the coefficients for different cations in NMA is $Bu_4N^+ > Pr_4N^+ > Et_4N^+ > Me_4N^+$. This is the order found for tetraalkylammonium fluorides[177] in water, but in water the coefficients of the chlorides and bromides show a reversal of the cation order[176, 177]. Nevertheless, since the differences in the coefficients for the cations in NMA become less pronounced as the anion size is increased, the similarity between the osmotic coefficients for those salts in NMA and water remains very marked. This similarity exists despite the "solvophobic interactions" which occur for tetraalkylammonium cations in water but not in NMA. The osmotic coefficients of alkali metal carboxylate salts were also found to follow essentially the same order in NMA and in water.

The osmotic coefficients do not appear to be very sensitive to solvent structural differences and, indeed, this also has been noted when results from salt solutions in water and deuterium oxide were compared[178].

f) Enthalpies and Entropies of Dilution

Falcone and Wood[179] have reported enthalpies of dilution in NMA for a large series of alkali metal halides as well as for a number of tetraalkylammonium halides (all measurements at 35 °C). These heats were expressed as excess enthalpies[180, 181] and values of the excess free energies calculated from the previously mentioned osmotic coefficients allowed calculation of the excess entropies.

The excess enthalpies of the alkali metal halides in NMA are of the same order of magnitude as the excess enthalpies of these same salts in water[179]. The specific anion appears to have little influence on the values in NMA but changing from one cation to another with the same halide ion has a large effect — the order being $Li^+ \simeq Na^+ > Cs^+ > K^+$. In water the cation effect is also large but the order is slightly different ($Li^+ > Na^+ \simeq K^+ > Cs^+$). This may reflect different changes in solvation number with changing cation size in the two solvents.

The results for the tetraalkylammonium salts in NMA differ greatly from the results in water[182], particularly at higher concentrations and also for solutions of the larger tetraalkylammonium ions as is shown in Table VI-3.

Interaction of the large tetraalkylammonium cations with water apparently leads to much more positive excess enthalpies than the interaction of the same ions in NMA. The excess enthalpy values for the smaller tetraalkylammonium ions do not appear to differ as markedly (or even in the same sense). The different tetraalkylammonium bromides in NMA show only slight and irregular changes with changing cation size. The excess entropies for the halide salts of the larger tetraalkylammonium cations tend toward much smaller positive values at high concentrations in NMA than at comparable concentrations in water[182].

Table VI-3. Excess enthalpies (kJ · mol⁻¹) of tetraalkylammonium halides in water and in NMA

Salt	0.3 Molal		0.8 Molal	
	H_2O	NMA	H_2O	NMA
Et$_4$NBr	−920	−100	−2090	−820
Pr$_4$NBr	−150	−570	+250	−1720
Bu$_4$NBr	+1880	−400	+7280	−1460
Pr$_4$NCl	+380	+130	+1320	−490
Pr$_4$NI	−430	−1660		

The heat of mixing in NMA from the six possible mixing experiments involving NaBr, NaI, Pr$_4$NBr and Pr$_4$NI (taken in pairs) have been reported by Falcone and Wood[183]. They have concluded that interactions occur between the iodide ions and the tetra-n-propylammonium ions in NMA solution which are comparable to the interactions between potassium ions and nitrate ions in aqueous solution.

g) Apparent Molal Volumes

Gopal, Siddiqi and Singh[184−186] have reported apparent molal volume (Φ_v) data for a variety of salts in NMA at 35 °C and for a more limited series of salts at several temperatures[184]. The apparent molal volumes appear to vary linearly with the square root of the concentration. The slopes of the Φ_v versus $\sqrt{C}$ plots are positive for the iodides of the smaller tetraalkylammonium ions but negative for the larger ($\geqslant$Bu$_4$N$^+$) tetraalkylammonium ions. This behavior essentially parallels that of the same salts in aqueous solution[187]. As solvophobic behavior in NMA is unlikely, it has been suggested[186] that the negative slopes are partially the result of solvent molecule and anion penetration between the organic groups of the tetraalkylammonium ions. The temperature dependence of Φ_v also has been discussed by Singh and co-workers[188].

In water[187], however, the slopes of the Φ_v versus $\sqrt{C}$ plots become positive for even these salts at very low solute concentrations. This is expected as a limiting positive slope is predicted by the Debye-Hückel equation for all salts in all solvents[189]. Attempts to demonstrate a change to positive slope for the larger tetraalkylammonium iodides in NMA have not been successful[186]. These attempts used a method of limited precision and cannot be considered to be conclusive.

More puzzling are the negative slopes found for the Φ_v versus $\sqrt{C}$ plots for certain simpler salts (KNO$_3$, KBr, NaCl, NaBr and NH$_4$Br) in NMA although several other similar salts give positive slopes[186]. Similar results have been found for salts in N-methylpropionamide[190]. Again no evidence has been found that lowering the concentration reverses the direction of the slopes. It is apparent that, without further experimental values, no attempt should be made to obtain partial molal volumes for any of these salts in NMA by extrapolation using concentration dependence of the apparent molal volumes.

h) Activity Coefficients from Gas Chromatographic Studies

Smiley[191] has measured gas-liquid retention volumes to obtain values for the activity coefficients at infinite dilution for eight different five-carbon hydrocarbons in NMA. The activity coefficients were determined at 40, 70, and 100 °C and, from the temperature dependence, values for the partial molar heats of solution were calculated.

Frost and Bittrich[192] have reported limiting activity coefficients of benzene and cyclohexane in NMA at 25 and 50 °C.

VII. Solvation

a) Acids and Bases in NMA

The behavior of acids and bases in NMA has been examined by several groups of researchers. Values for the pK_a of acids determined in these potentiometric and conductometric studies are summarized in Table VII-1. The measurements have not been carried out at any single temperature and one extensive study reports values

Table VII-1. pK_a values for acids in NMA

Acid	t °C	Method	pK_a	Ref.
HCl	40	c	s	115)
	35	c	s	105)
	?	p	s	140)
HBr	35	c	s	105)
	37	p[1]	s	144)
HI	35	c	s	105)
HClO$_4$	?	pol	s	140)
	40	c	s	195)
2,4,6-Trinitrophenol	40	c	s	115)
p-Toluenesulfonic	40	p	s	141)
H$_2$SO$_4$	37	p[1]	s, 2.8	144)
Trifluoroacetic	30	c	1.55	127)
	50	c	1.54	127)
Sulfanilic	40	c	1.66	16)
Metanilic	40	c	2.09	16)
Pyridinium ion	37	p[2]	2(?)[2]	144)
Dichloroacetic	30	c	2.90	127)
	37	p[1]	3.1	144)
	40	c	2.94	196)
	50	c	2.89	127)
Analinium ion	37	p[1]	3	144)
2,4-Dinitrophenol	40	c	3.83	16)
Salicylic	40	c	3.85	140)
Malonic	37	p[1]	4.7	144)

Table VII-1 (continued)

Acid	t °C	Method	pK$_a$	Ref.
Chloroacetic	30	c	4.97	127)
	37	p[1]	5	144)
	40	c	4.76[3]	196)
Benzoic	?	c	6.7	104)
	37	p[1]	6.9	144)
	40	c	6.42	16)
2,4,6-Trichlorophenol	40	c	6.60	16)
Cinnamic	40	c	7.16	16)
Acetic	37	p[1]	7.5	144)
	40	c	7.16	16)
Propionic	40	c	7.32	16)
Butyric	40	c	7.34	16)
Valeric	40	c	7.34	16)
Trimethylacetic	40	c	7.60	16)
4-Nitrophenol	40	c	8.23	16)
	37	p[1]	8.8	144)
Benzylammonium ion	37	p[1]	8.6	144)
2,4-Dichlorophenol	40	c	9.03[3]	16)
Diethylammonium ion	37	p[1]	9.3	144)
Piperidinium ion	37	p[1]	9.3(?)[2]	144)
n-Butylammonium ion	37	p[1]	9.5	144)
Methylammonium ion	37	p[1]	12.2	144)
Phenol	37	p[1]	13.1	144)
Dihydrobenzoquinone	37	p[1]	13.5, 15.3	144)

c = conductometric method; p = potentiometric method; pol = polarographic method;

[1]) Designates determination of pK$_a$ with 0.3 M LiClO$_4$ as added electrolyte.

[2]) The pK$_a$ value of 2 for the pyridinium ion in NMA is an approximate value deduced from Figure 5 of Ref.[144]. The value of 9.3 for the pK$_a$ for the acid-base couple C$_5$H$_6$N$^+$/C$_5$H$_5$N in the table of Ref.[144] probably should be attributed to the piperidinium/piperidine couple (again see Fig. 5 of Ref.[144]). Reynaud[197] has reported the pK$_a$ of the pyridinium ion in H$_2$O/NMA mixtures. The pK$_a$ of this ion tends to decrease with increasing NMA concentration and has a value of 2.48 at 80% NMA by weight.

[3]) This value depends on an estimated Λ$_0$ value.

determined with 0.3 M LiClO$_4$ as an added electrolyte[144]. However, the work of Fialkov and Suprunenko[127] suggests that, as is the case for aqueous solution[193], the dissociation constants for carboxylic acids in NMA vary little with temperature. Further, considering the high dielectric constant of NMA, values for the acid pK$_a$'s, determined in an NMA medium of even moderately high ionic strength, would not be expected to be changed greatly from the "zero ionic strength" values.

HCl, HBr, HI and HClO$_4$ have been shown to behave as strong acids in NMA solution. The dissociation of the first proton of H$_2$SO$_4$ has been found to be essentially complete in NMA while dissociation of the second proton is less favored in NMA[140, 144] than in water[194]. NaOH and Bu$_4$NOH are strong bases in NMA solutions[144]. NMA containing 0.05 M water shows an accessible range of 18.3 pH units between 1 M HBr and 1 M NaOH[144].

Simple carboxylic acids are markedly weaker in NMA than in water[194]. The more weakly acidic of the phenols in water are also weaker in NMA but to a lesser degree. However, those phenols which are somewhat stronger acids in water ($pK_a < 4$) appear to be at least equally strong acids in NMA [e.g., 2,4-dinitrophenol and 2,4,6-trinitrophenol (or picric acid)].

The pK_a's of the carboxylic acids and of the phenols in NMA separately show rough linear correlations with the pK_a's of the same acids in aqueous solution[16]. In both cases the slope [$\Delta pK_a(NMA)/\Delta pK_a(H_2O)$] is greater than one.

In general (except for the methylammonium ion), those cationic acids which have been investigated are more acidic in NMA than in water[194]. The two amino-benzenesulfonic acids also are stronger in NMA than in water[194]. These two acids may be present as zwitterions in NMA as sulfanilic acid is in water[198]. If this is the case, these acids might be expected to behave in a fashion similar to the simple cationic acids[16]. No simple linear correlation between $pK_a(NMA)$ and $pK_a(H_2O)$ is apparent for these acids.

The higher dielectric constant of NMA, compared to water, might be expected to promote dissociation of acids but studies in other nonaqueous solvents and in mixed solvents indicate that the dielectric constant is seldom the predominant factor controlling acid dissociation processes[199]. Instead, whether a particular acid is stronger in one solvent or another, will likely be quite dependent on the relative solvation of the acid, of the proton and of the conjugate base in the two solvents.

In the case of NMA, Pucci[141] has shown that formation of the solvated H^+ ion is more favored thermodynamically in NMA than it is in water; i.e., NMA is a stronger base than water. Indeed it appears that cations, in general, are better solvated in NMA than in water[141]. Anions, on the other hand, are destabilized on transfer from water to NMA and destabilization of anions in NMA can be larger than stabilization of the proton. Thus, it is quite probable that the weakness of the carboxylic acids and of most of the phenols in NMA as compared to water is the result of the corresponding carboxylate ions and phenolate ions being very poorly solvated in NMA. It is also not surprising, then, that the cationic acids are generally as strong or stronger in NMA as in water, since no anion is produced from such a dissociation process. The acid strength of the cationic acid is probably mainly a function of the relative solvation of the proton and the cationic species itself, although solvation of the uncharged conjugate base is also likely to be important. Trends in acid strengths in NMA are similar to those which have been noted for acids in formamide[200] and acetamide[201].

b) Diffusion

Williams et al.[92] have investigated the diffusion of the sodium ion in NMA solutions of NaCl at 40 °C. The values of the diffusion coefficient for the Na^+ ion were found to be equal to approximately half the value of the self-diffusion coefficient for pure NMA. This suggests a solvation which is equivalent to three or four molecules of NMA coordinated to the sodium ion.

c) Viscosity Studies on Solutions of Salts in NMA

In one of the earlier papers on the use of NMA as a solvent, it was pointed out[11] that, as the concentration of certain salts was increased, the viscosities of the NMA solutions were also greatly increased (*e.g.,* at 40 °C the viscosity of a 0.4 molal NaI solution is approximately 30% higher than the viscosity of pure NMA). Such large increases in solution viscosity usually have been attributed to solvation of the ions of the added electrolyte[202].

Gopal and Rastogi[84] have determined the temperature and concentration dependence of the viscosities of solutions of a number of salts (mainly tetraalkylammonium iodides) in NMA. They interpreted their results in terms of the Jones-Dole equation[203]: $\eta = \eta_0(1 + A\sqrt{C} + BC)$. The value of B was calculated to be positive for all of the salts examined in NMA. For LiCl and KI the value of B was found to decrease with increasing temperature. A similar trend can be calculated for the temperature dependence of B for KBr and NaI[204]. On the other hand, the B coefficients for the tetraalkylammonium iodides increased with increasing temperature.

In aqueous solutions the viscosity B coefficients have been found to be additive properties of the constituent ions within a reasonable degree of accuracy, and are strongly correlated with the entropies of solution of the ions[202]. Ions which are apparently structure-breaking tend to have B values which increase with an increase in temperature. This is the result of the structure-breaking ability of the ions becoming less in a relative sense as the thermal agitation of the solvent molecules decreases the solvent structure. Conversely, for structure-making ions the B coefficients tend to decrease with increasing temperature.

Gopal and Rastogi[84] attempted to divide the viscosity B values for salts in N-methylpropionamide and to assign a separate B coefficient to each of a series of cations and anions in this solvent. The viscosity data for NMA solutions were then discussed by analogy. It was concluded that the B values for the alkali metal ions decrease with increasing temperature indicating a net reinforcement of solvent structure by these ions. On the other hand, the B values for the halide ions increase with increasing temperature, suggesting that these ions either have structure-breaking properties or else they become more solvated with increasing temperature or both. As noted earlier, heat capacity measurements in amide solvents suggest that such solvent-breaking properites are much smaller than are the effects of ions like Cs^+ or Br^- on the structure of water[170].

The B values for the tetraalkylammonium ions (particularly the lower molecular weight homologues) were found, by Gopal and Rastogi[84], also to increase with increasing temperature. Falcone[205], however, has re-analyzed the data of these workers and concludes that the B values change very little or actually decrease slightly with increasing temperature. It appears that these ions have only a slight effect on the structure of the NMA solvent.

d) Walden Products

Dawson and co-workers[115] examined the Walden products for a number of ions at 40 °C. They found the products in NMA to be higher than those in water for the large organic ions and considerably lower than those in water for most inorganic ions studied in NMA. Singh et al.[75, 121, 206] have examined the Walden products at several temperatures for a variety of ions in NMA using the transport number measurements of Gopal and Bhatnagar[135]. The values of the Walden products for the $Li^{+b)}$, Na^+ and K^+ ions were found to change irregularly with temperature. This has been interpreted as reflecting that these ions are neither structure-promoting nor structure-breaking in NMA (although it should be noted that $\lambda_0\eta$ for Li^+ in water is essentially temperature independent)[207]. The irregularity might also merely reflect errors in the Λ_0 values (derived in some cases from the interpolation of a curve of Λ_0 versus temperature for a single salt) or errors in the transport number values which, as previously noted, may be suspect although it would be expected, at least, that the relative values obtained for the transport numbers at different temperatures would be correct. The Walden products for the larger alkali metal and halide ions clearly decrease with increasing temperature, as do those for the tetraalkylammonium ions. Singh has suggested[206] that this infers structure-breaking by these ions in NMA for the same reason noted for the temperature dependence of the viscosity of salt solutions in NMA and, hence, that these ions are solvated in NMA.

e) Coordination of NMA and Solvates

One indication of the solvating ability of NMA is the large number of solvates which are known. Such solvates have been noted, not only for many salts containing multi-charged cations [e.g., $AlCl_3 \cdot 10$ NMA[159], $PbBr_2 \cdot 2$ NMA[208], $Pd(NMA)_4(BF_4)_2$][209] but also for alkali metal salts (e.g., $LiBr \cdot 4$ NMA)[210] and for simple acids (e.g., $HCl \cdot 2$ NMA)[159].

The crystal structure of $LiCl \cdot 4$ NMA[210] has been determined and it was found that the NMA molecules are coordinated through the oxygen to the Li^+ ions, the nitrogen atoms being directed towards the Cl^- ions. This suggests, but of course does not prove, that in NMA solutions of lithium salts, the Li^+ ions are solvated by oxygen coordinated NMA molecules. This is in agreement with MO calculations[211] which indicate that such coordination would be energetically favorable (but much less so for the other alkali metal ions), and with infrared spectroscopic data[212] for NMA solutions of LiBr. The crystal structures of $NaClO_4 \cdot 2$ NMA[213] and of $KI \cdot KI_3 \cdot 6$ NMA[214] also have been determined. These too suggest oxygen coordination of NMA to the cation and nitrogen coordination to the anion. In the case of $KI \cdot KI_3 \cdot 6$ NMA, the I_3^- ion is markedly more distant from the NMA nitrogen atoms than is the I^- ion. Rao and co-workers[215] have used infrared spectroscopy to deduce that the NMA is coordinated through the oxygen to the magnesium ion in $Mg(NMA)_6Cl_2$.

Baron and de Lozé[212] have studied the interaction of several salts with NMA using IR and Raman spectroscopy. It was found that, in very concentrated solutions, even tetra-n-butylammonium bromide caused shifting of the amide carbonyl bands

---

b) The value for $\lambda_0\eta(Li^+, 35\,°C)$ in Table 4 of Ref.[121] is 0.192 $ohm^{-1} \cdot cm^2 \cdot eq^{-1} \cdot P$ if the reasonable value of $\lambda_0(Li^+, 35\,°C)$ from the same reference (Table 1) is used.

while for $NaClO_4$ and $LiClO_4$ there was evidence that two cations can interact with a single carbonyl oxygen of NMA, the interactions with Li^+ being stronger than those with Na^+. No evidence was found for coordination of two Ba^{2+} ions to a single NMA molecule in very concentrated NMA solutions of $Ba(ClO_4)_2$. The Raman spectra of NMA mixtures with perchlorate salts suggested that the NMA-salt interactions were accompanied by formation of an amount of free (not self-associated) NMA.

While there are many indications that oxygen coordinated solvation of cations in NMA is predominant, there have been suggestions that nitrogen coordination solvation of cations also may occur[216, 217]. Much of the evidence for this is based on the interpretation of vibrational spectral data for solid solvates and these data have shown a remarkable flexibility toward interpretation[217, 218]. It is quite possible that the type of coordination may depend strongly on the particular ion involved.

f) Behavior of Bromine in NMA

As in many other organic solvents[219], the conductance of solutions of bromine in NMA has been found to be quite high[220]. Such solutions have been examined by spectrophotometry, cyclic voltammetry, and acid-base titration[146]. It was determined that both Br^- and Br_3^- are present in such solutions, while the predominant cation is H^+. As the concentration of bromine in NMA is increased the concentration of $Br^+(NMA)$ formed also becomes significant. The following equilibria were proposed[146] to explain these results:

1. $NMA + Br_2 = Br^- + Br^+(NMA)$
2. $Br^+(NMA) = H^+ + CH_3COBrNCH_3$
3. $Br_3^- = Br_2 + Br^-$

g) Micelle Formation

Gopal and Singh have shown that the conductance curves for the sodium salts of large carboxylate ions (oleate, stearate, palmitate, deoxycholate) as well as for hexadecyltrimethylammonium bromide in NMA strongly resemble similar curves which have been described as indicative of micelle formation in aqueous solution[221, 222]. At 35 °C the critical micelle concentrations for these salts were found to range from 0.7 to 0.9×10^{-4} M. Refractive index studies[223] have confirmed the formation of micelles (although suggesting critical micelle concentrations which are larger by a factor of about two). Similar results have been found for these same salts in formamide[223, 224]. These workers have suggested that, since they found no evidence of micelle formation in N,N-dimethylformamide and N,N-dimethylacetamide, micelle formation generally is favored by a solvent of high dielectric constant which is also capable of strong hydrogen bonding.

The conductance data of Wilhoit[225] for octadecyltrimethylammonium nitrate, octadecylsulfate, and iodide show no indication of micelle formation at low concentrations (down to $c = 9.48 \times 10^{-5}$ M, the phoreograms are linear) although this may simply be a result of the critical micelle concentration being at even lower concentrations. Also somewhat puzzling are the phoreograms for sodium deoxycholate in DMF and DMA[221] which suggest that Λ_0 for the salt is very similar to the value of the limiting ionic equivalent conductance of Na^+ alone[226, 227]. Clearly more work is necessary to enlarge our understanding of the behavior of dilute solutions of large ions in NMA.

h) Complexation Constants in NMA

Several groups of workers have studied the ability of NMA molecules and of anions to compete for sites in the first coordination sphere about various metal cations in NMA solution. The values of the experimentally determined formation constants are shown in Table VII-2.

Table VII-2. Complex Formation Constants in NMA

Cation	Anion	$t\,°C$	β_1	β_2	β_3	β_4	Ref.
Hg^{2+}	SCN^-	40		14.6	18.4	20.3	141)
Hg^{2+}	I^-	37				26.5	148)
Ag^+	SCN^-	40			10.7		141)
Fe^{3+}	Cl^-	30	2.8	4.0			228)
Sn^{2+}	Br^-	35	1)(1.96	4.74	6.84	8.63)	149)
			(2.08	4.41	7.14	$-$)	

$\beta = \log_{10}$ formation constant.
1) The maximum number of Br^- ions complexed per Sn^{2+} ion is uncertain.

Drago et al.[229] have indicated that identical UV-visible spectra were obtained for dilute solutions of $FeCl_3$, $Et_4N(FeCl_4)$ and $Fe(NMA)_6(ClO_4)_2$ in NMA solution. This suggests that in dilute solution the predominant ionic species is $Fe(NMA)_x^{3+}$. The values reported by Reynolds and Weiss[228] for the formation of $FeCl_2^+$ and $FeCl^{2+}$ were obtained from experiments using higher chloride concentrations. These workers also considered the formation of a deprotonated solvolyzed species $Fe(HS)_y(S)^{2+}$.

Thompson[149] was unable to distinguish potentiometrically if tin(II) species involving four Br^- ions or only three Br^- ions were formed at high Br^- ion concentrations. Regardless, the values for the formation constants are markedly lower than values similarly obtained for tin(II)/bromide complexes in N,N-dimethylformamide reflecting the stronger solvating ability of NMA.

The formation constants of the Hg^{2+} salts appear quite similar to the values in water. This is probably the result of the Hg^{2+} ion being more readily solvated in NMA than in water while the reverse will tend to be true for solvation of anions[141]. Balyatinskaya[230] has used the results of Pucci *et al.*[141] to show that $\log_{10} {}_{H_2O}\gamma^{NMA}$ is negative for $Hg(SCN)_x^{(2-x)+}$ ($x = 2, 3, 4$).

i) Surface Tension of Salt Solutions

Gopal and Bhatnagar[231] have examined the changes in surface tension of NMA solutions of tetraalkylammonium iodides with changing salt concentration. The relative surface tension rapidly increases with increasing salt concentration (up to about 0.05 M). This is in contrast to aqueous solutions of these solutes where the relative surface tension initially decreases and then increases. This difference in behavior was attributed to the lack of solvophobic interactions of the tetraalkylammonium ions in NMA.

j) Proteins and Polyaminoacids as Solutes in NMA

The solubility of a number of proteins in NMA was investigated by Rees and Singer[232]. Of these insulin and β-lactoglobulin were found to be soluble to at least one gram per liter while zein is soluble to at least ten grams per liter[233]. An ORD study of zein in NMA[233] led to the conclusion that the helicity (28 °C) is 42%. While this is greater than was found in any other of the several solvents tested, it was established that the helicity was not a simple function of the solvent dielectric constant.

Helix formation of poly(L-glutamic acid) has been studied by ORD[234], by potentiometric titration[235] and by viscosity methods[236]. Thermodynamic parameters for the coil-helix transition of the uncharged polymer have been determined and it has been found[235] that the helix is an order of magnitude more stable in NMA than in water.

VIII. Reactions in NMA

Several examples of complex formation in NMA of the type, $M^{n+} + a\,Y^- = MY_a^{(n-a)}$, have been discussed previously. While these might easily be classified as solvolysis reactions, such examples will not be considered further.

Eckstrom and co-workers[237] have reported a study of the solvolysis of *cis* and *trans* $[Co(en)_2X_2]X$ ($X = Cl^-, Br^-$) in NMA. The chloro and bromo complexes appeared to react similarly with a single halide ion in the first coordination sphere of the cobalt ion being replaced by an NMA molecule. *Cis* and *trans* isomers apparently gave an identical product. The rates of the reactions were similar to those found for the solvolysis of the same complexes in other solvents. The reaction products were not characterized. Neither the *cis* nor the *trans* complex $[Co(en)_2(NO_2)_2]NO_3$ appeared to solvolyze readily in NMA.

The solvolysis reactions of *cis* and *trans* *t*-butylcyclohexyl tosylates also have been studied in NMA[238]. In both cases the primary product was 4-*t*-butylcyclohexene (no other cyclohexene products were detected) and much smaller quantities of cyclohexyl acetates and cyclohexanols were also recovered. The reaction rate was first order with respect to the tosylates. Similar to results of studies of the reaction in other solvents, the *cis*-tosylate was solvolyzed more readily than was the *trans* compound. The stereochemical distribution of the minor products was significantly altered by small amounts of water ($<1\%$) added to the NMA solvent.

Horclois *et al.*[239] have suggested the use of NMA as a solvent for the reaction of thiosemicarbazide with carbon disulfide to produce 2-amino-5-mercapto-1,3,4-thiadiazole. The reaction proceeds at a lower pressure (atmospheric) and at a lower temperature than the comparable reaction in water. Campbell[240] has described the polymerization of acrylonitrile by BF_3 in amide solvents including NMA.

Dawson *et al.*[159] have reported a very cursory study of several oxidation-reduction and metathetical reactions carried out in NMA. It was generally found that the results indicated behavior parallel to that found for the reactions in water.

The reaction of sodium phenoxide with several different alkyl iodides in NMA was studied in more detail[118]. The kinetics were again found to be similar to those for the same reaction in a markedly different solvent (in this case, ethanol)[241] although the activation energy is about $10 \, kJ \cdot mol^{-1}$ lower for the reaction in NMA. The activation energy in NMA is, however, higher than was found for the reaction in dimethylacetamide-NMA mixtures[118]. It is probable that these differences are mainly a function of the differing abilities of solvents to solvate the phenoxide ion and the activated complex and are not primarily related to the differences in dielectric constants.

Hall[242] noted that the reaction of NMA with an acyl chloride liberated free chloride ions at a much slower rate than the comparable reaction in N,N-dimethylacetamide or in N,N-dimethylformamide.

Yoneda *et al.*[243] have investigated the reaction between alkyl halides and thiocyanate ion in a large number of solvents including NMA. While the rate of reaction in NMA was greater than in many of the solvents tested (*e.g.,* formamide), the rate of product formation was still slower than in several solvents, particularly the N,N-dimethylamides and DMSO. Smiley[244] has suggested that the reaction of dimethyl sulfate with thiocyanate ion in NMA proceeds by two different pathways. Dimethyl sulfate can react initially with the solvent to form an ionic intermediate which then reacts with the thiocyanate ion. Dimethyl sulfate also can react directly with the thiocyanate ion.

The decomposition of trichloroacetate ion in NMA has been found[119] to follow pseudo-first order kinetics. The activation energy for the reaction was found to be intermediate between Verhoek's values for the activation energy of the decomposition in ethanol and in aniline[245].

In all of the above reactions, the high dielectric constant of NMA has been found to have a minimal effect on the reaction rates. This might not be the case if both reactants were highly charged species. Certainly, even in such a case, the ability of a particular solvent to solvate a given ion or transition state would be expected to remain important. Unfortunately, in those studies which have been reported of reactions between charged reactant species in mixed solvents containing a large mole fraction of NMA[246, 247], incomplete information concerning complexation constants has frustrated attempts at detailed analysis of the kinetic data.

IX. Other Studies

a) Biological Effects of NMA

The relatively low toxicity of NMA is similar to the toxicities of several other N-alkyl and N,N-dialkyl substituted carboxylic acid amides and has been the subject of several studies. There is general agreement that the lethal dose (LD_{50}) for rats is of the order of $4 \text{ g} \cdot \text{kg}^{-1}$ [248–250]. Large sublethal doses induce reversible diabetogenic effects[248, 251–253] apparently by rendering inactive both secreted and injected insulin[248]. Doses greater than $0.8 \text{ g} \cdot \text{kg}^{-1} \cdot \text{day}^{-1}$ are cumulative while smaller daily doses are not toxic within a period of six to eight months (although normal weight gain is apparently affected)[249]. Caryoclastic action of NMA as a result of large (0.8 to $8 \text{ g} \cdot \text{kg}^{-1}$) doses also has been noted[254]. The lethal dose found for mice is similar to that for rats although their apparent reaction to the NMA was somewhat different[249]. Studies[250, 255] also have indicated that even 23% of the lethal dose (LD_{50}) of NMA has a teratogenic effect in rats.

b) Miscellaneous Uses of NMA as a Solvent

Several further uses of NMA as a solvent have been described. Dawson and Oei[256] have reported the preparation of CuS and As_2S_3 sols in NMA. The As_2S_3 sol was found to be much more stable in NMA than in water — probably primarily because of the higher dielectric constant of NMA. The CuS sol was found to go into solution within less than two weeks. Patents[257, 258] have been obtained for the use of NMA as a solvent in the preparation of photographic materials. Papers[259, 260] have appeared which discuss the potential use of NMA solutions of $AgNO_3$ and $PdCl_2$ as stationary phases for gas chromatography separations of hexenes. NMA has been patented[261] as a paint remover for latex paints.

X. References

[1] Leader, G. R., Gormley, J. F.: J. Amer. Chem. Soc. *73*, 5731 (1951)
[2] Charlot, G., Trémillon, B.: Chemical reactions in solvents and melts (1963); translated by Harvey, P. J. J., Oxford: Pergamon Press 1969
[3] Waddington, T. C.: Non-aqueous solvent systems. New York: Academic Press 1965
[4] Dawson, L. R.: in: Chemie in nichtwäßrigen ionisierenden Lösungsmitteln (ed. G. Jander, H. Spandau, C. C. Addison). Vol. 4, Part 5. New York: Interscience 1963, p. 257

5) Vaughn, J. W.: in: The Chemistry of Non-Aqueous Solvents (ed. J. J. Lagowski), Vol. 2, Chap. 5. New York: Academic Press 1967, p. 225

6) Reid, D. S., Vincent, C. A.: J. Electroanal. Chem. *18*, 427 (1968)

7) Knecht, L. A.: Pure Appl. Chem. *27*, 281 (1971)

8) Knecht, L. A., Kolthoff, I. M.: Inorg. Chem. *1*, 195 (1962)

9) Miyazawa, T., Shimanouchi, T., Mizushima, S.: J. Chem. Phys. *24*, 408 (1956)

10) Berger, A., Loewenstein, A., Meiboom, S.: J. Amer. Chem. Soc. *81*, 62 (1959)

11) Dawson, L. R., Sears, P. G., Graves, R. H.: J. Amer. Chem. Soc. *77*, 1986 (1955)

12) Rudnick, K., Geipel, H.: Patent 79, 724, German Democratic Republic (1971)

13) Bonner, O. D., Jordan, C. F., Bunzl, K. W.: J. Phys. Chem. *68*, 2450 (1964)

14) Sears, P. G., Stoeckinger, T. M., Dawson, L. R.: J. Chem. Eng. Data *16*, 220 (1971)

15) Knecht, L. A.: personal communication (1965)

16) Edmonson, K. A.: Ph. D. Dissertation, University of Kentucky, Lexington, Ky., 1967

17) Bonner, O. D., Woolsey, G. B.: J. Phys. Chem. *75*, 2879 (1971)

18) Kimura, M., Aoki, M.: Bull. Chem. Soc. Japan *26*, 429 (1953)

19) Kitano, M., Fukuyama, T., Kuchitsu, K.: Bull. Chem. Soc. Japan *46*, 384 (1973)

20) Gilpin, J. A.: Anal. Chem. *31*, 935 (1959)

21) Tereshkovich, M. O., Volkova, S. A., Bereznitskii, Z. S., Topil'skii, G. S.: Kompleksoobrazov., Mezhmol Vzaimodeistvie Soosazhd. Nekot. Sist. 15 (1970)

22) Higasi, K., Omura, I., Baba, H.: Nature *178*, 652 (1956)

23) Meighan, R. M., Cole, R. H.: J. Phys. Chem. *68*, 503 (1964)

24) Nakagawa, I.: Nippon Kagaku Zasshi *79*, 1353 (1958); Chem. Abst. *53*, 7699a (1959); Chemisches Zentralblatt *130*, 17000 (1959)

25) Mizushima, S., Simanouti, T., Nagakura, S., Kuratani, K., Tsuboi, M., Baba, H., Fujioka, O.: J. Amer. Chem. Soc. *72*, 3490 (1950)

26) Lin, R.-Y., Dannhauser, W.: J. Phys. Chem. *67*, 1805 (1963)

27) Kaya, K., Nagakura, S.: Theor. Chim. Acta *7*, 117 (1967)

28) Kaya, K., Nagakura, S.: Theor. Chim. Acta *7*, 124 (1967)

29) Nielsen, E. B., Schellman, J. A.: J. Phys. Chem. *71*, 2297 (1967)

30) Jones, R. L.: J. Mol. Spectrosc. *11*, 411 (1963)

31) Hallam, H. E., Jones, C. M.: Trans. Faraday Soc. *65*, 2607 (1969)

32) Jones, R. L.: Spectrochim. Acta *23A*, 1745 (1967)

33) Russell, R. A., Thompson, H. W.: Spectrochim. Acta *8*, 138 (1956)

34) Miyazawa, T.: J. Mol. Spectrosc. *4*, 155 (1960)

35) Barker, R. H., Boudreaux, G. J.: Spectrochim. Acta *23A*, 727 (1967)

36) Liler, M.: J. Chem. Soc. Perkin Trans. *II*, 720 (1972)

37) Klotz, I. M., Franzen, J. S.: J. Amer. Chem. Soc. *84*, 3461 (1962)

38) Graham, L. L., Chang, C. Y.: J. Phys. Chem. *75*, 776 (1971)

39) Bhaskar, K. R., Rao, C. N. R.: Biochim. Biophys. Acta *136*, 561 (1967)

40) Lindheimer, M., Étienne, G., Brun, B.: J. Chim. Phys. *69*, 829 (1972)

41) Rassing, J.: Ber. Bunsenges. Phys. Chem. *75*, 334 (1971)

42) Longsworth, L. G.: J. Colloid Interfac. Sci. *22*, 3 (1966)

43) Gopal, R., Rizvi, S. A.: J. Indian Chem. Soc. *47*, 287 (1970)

44) Grigsby, R. D., Christian, S. D., Affsprung, H. E.: J. Phys. Chem. *72*, 2465 (1968)

45) Albers, R. J., Swanson, A. B., Kresheck, G. C.: J. Amer. Chem. Soc. *93*, 7075 (1971)

46) Løwenstein, H., Lassen, H., Hvidt, A.: Acta Chem. Scand. *24*, 1687 (1970)

47) Kresheck, G. C., Kierleber, D., Albers, R. J.: J. Amer. Chem. Soc. *94*, 8889 (1972)

48) Shipman, L. L., Christofferson, R. E.: J. Amer. Chem. Soc. *95*, 1408 (1973)

49) Ramachandran, G. N., Lakshminarayanan, A. V., Kolaskar, A. S.: Biochim. Biophys. Acta *303*, 8 (1973)

50) Murthy, A. S. N., Rao, K. G., Rao, C. N. R.: J. Amer. Chem. Soc. *92*, 3544 (1970)

51) Momany, F. A., McGuire, R. F., Yan, J. F., Scheraga, H. A.: J. Phys. Chem. *74*, 2424 (1970)

52) Perricaudet, M., Pullman, A.: Int. J. Peptide Protein Research *5*, 99 (1973)

53) Kolaskar, A. S., Lakshminarayanan, A. V., Sarathy, K. P., Sasisekharan, V.: Biopolymers *14*, 1081 (1975)

54) Ramaprasad, S., Kellerhals, H. P., Kunwar, A. C., Khetrapal, C. L.: Mol. Cryst. Liq. Cryst. *34*, 19 (1976)

55) Costain, C. C., Dowling, J. M.: J. Chem. Phys. *32*, 158 (1960)

56) Reeves, L. W., Riveros, J. M., Spragg, R. A., Vanin, J. A.: Mol. Phys. *25*, 9 (1973)

57) Drakenberg, T., Forsén, S.: J. Chem. Soc. *D*, 1404 (1971)

58) Rabinovitz, M., Pines, A.: J. Amer. Chem. Soc. *91*, 1585 (1969)

59) Yan, J. F., Momany, F. A., Hoffmann, R., Scheraga, H. A.: J. Phys. Chem. *74*, 420 (1970)

60) Katz, J. L., Post, B.: Acta Cryst. *13*, 624 (1960)

61) Dentini, M., De Santis, P., Morosetti, S., Piantanida, P.: Z. Krist. *136*, 305 (1972)

62) Kojima, T., Kawabe, K.: Oyo Butsuri *42*, 9 (1973)

63) Suga, H., Nakatsuka, K., Shinoda, T., Saki, S.: Nippon Kagaku Zasshi *82*, 29 (1961); Chem. Abst. *55*, 11952a (1961)

64) Bradbury, E. M., Elliott, A.: Spectrochim. Acta *19*, 995 (1963)

65) Itoh, K., Shimanouchi, T.: Biopolymers *5*, 921 (1967)

66) Pivcová, H., Schneider, B., Štokr, J., Jakeš, J.: Coll. Czech. Chem. Commun. *29*, 2436 (1964)

67) Schneider, B., Hoření, A., Pivcová, H., Honzl, J.: Coll. Czech. Chem. Commun. *30*, 2196 (1965)

68) Pivcová, H., Schneider, B., Štokr, J.: Coll. Czech. Chem. Commun. *30*, 2215 (1965)

69) Jakeš, J., Schneider, B.: Coll. Czech. Chem. Commun. *33*, 643 (1968)

70) Aihara, A.: J. Chem. Soc. Japan Pure Chem. Sect. *73*, 855 (1952)

71) Fillaux, F., De Lozé, C.: J. Chim. Phys. *73*, 1004 (1976)

72) Fillaux, F., De Lozé, C.: J. Chim. Phys. *73*, 1010 (1976)

73) Fillaux, F., De Lozé, C.: Chem. Phys. Lett. *39*, 547 (1976)

74) Wood, R. H., Wicker, R. K., II, Kreis, R. W.: J. Phys. Chem. *75*, 2313 (1971)

75) Singh, R. D., Rastogi, P. P., Gopal, R.: Can. J. Chem. *46*, 3525 (1968)

76) French, C. M., Glover, K. H.: Trans. Faraday Soc. *51*, 1427 (1955)

77) Weeda, L., Somsen, G.: Rec. Trav. Chim. Pays-Bas *86*, 263 (1967)

78) Kortüm, G., Biedersee, H. V.: Chem. Ing. Tech. *42*, 552 (1970)

79) Manczinger, J., Kortüm, G.: Z. Phys. Chem. (NF) *95*, 177 (1975)

80) Gopal, R., Rizvi, S. A.: J. Indian Chem. Soc. *45*, 13 (1968)

81) Thompson, G. W.: Chem. Rev. *38*, 1 (1946)

82) Dawson, L. R., Zuber, W. H., Jr., Eckstrom, H. C.: J. Phys. Chem. *69*, 1335 (1965)

83) Gopal, R., Rizvi, S. A.: J. Indian Chem. Soc. *43*, 179 (1966)

84) Gopal, R., Rastogi, P. P.: Z. Phys. Chem. (NF) *69*, 1 (1970)

85) Paul, R. C., Banait, J. S., Singla, J. P., Narula, S. P.: Z. Phys. Chem. (NF) *88*, 90 (1974)

86) Johari, G. P., Tewari, P. H.: J. Phys. Chem. *70*, 197 (1966)

87) Kortüm, G., Hebestreit, C.: J. Phys. Chem. (NF) *95*, 65 (1975)

88) Casteel, J. F., Amis, E. S.: J. Chem. Eng. Data *19*, 121 (1974)

89) Assarsson, P., Eirich, F. R.: J. Phys. Chem. *72*, 2710 (1968)

90) Basov, V. P., Karapetyan, Yu. A., Krysenko, A. D.: Russ. J. Phys. Chem. *47*, 678 (1973)

91) Casteel, J. F., Sears, P. G.: J. Chem. Eng. Data *20*, 10 (1975)

92) Williams, W. D., Ellard, J. A., Dawson, L. R.: J. Amer. Chem. Soc. *79*, 4652 (1957)

93) Bass, S. J., Nathan, W. I., Meighan, R. M., Cole, R. H.: J. Phys. Chem. *68*, 509 (1964)

94) Lutskii, A. E., Mikhailenko, S. A.: J. Struct. Chem. *4*, 323 (1963)

95) Hovermale, R. A., Sears, P. G., Plucknett, W. K.: J. Chem. Eng. Data *8*, 490 (1963)

96) Itoh, K., Sato, H., Takahashi, H., Higasi, K.: Bull. Chem. Soc. Japan *49*, 329 (1976)

97) Kirkwood, J. G.: J. Chem. Phys. *7*, 911 (1939)

98) Fröhlich, H.: Theory of Dielectrics, p. 49. New York: Oxford University Press (1949)

99) Lustkii, A. E., Mikhailenko, S. A.: Russ. J. Phys. Chem. *38*, 775 (1964)

100) Oei, D. G.: unpublished results, University of Kentucky (1959)

101) Verma, R. K., Yaseen, M., Aggarwal, J. S.: Indian J. Technol. *4*, 139 (1966)

102) Grunberg, L., Nissan, A. H.: Trans. Faraday Soc. *45*, 125 (1949)

103) Kreis, R. W., Wood, R. H.: J. Chem. Thermodynam. *1*, 523 (1969)

104) Wicker, R. K., II.: Ph. D. Dissertation, University of Delaware, Newark, Delaware (1966)

105) Kortüm, G., Hebestreit, C.: Z. Phys. Chem. (NF) *93*, 235 (1974)
106) Ramiah, K. V., Chalapathi, V. V.: Proc. Indian Acad. Sci. *A60*, 242 (1964)
107) Momii, R. K., Urry, D. W.: Macromolecules *1*, 372 (1968)
108) Kamei, H.: Bull. Chem. Soc. Japan *38*, 1212 (1965)
109) Shimokawa, S., Sohma, J., Itoh, M.: Bull. Chem. Soc. Japan *40*, 693 (1967)
110) LaPlanche, L. A., Rogers, M. T.: J. Amer. Chem. Soc. *86*, 337 (1964)
111) Miyazawa, T.: Bull. Chem. Soc. Japan *34*, 691 (1961)
112) Bonner, O. D., Kim, S. J.: J. Chem. Thermodynam. *2*, 63 (1970)
113) Wood, R. H., DeLaney, D. E.: J. Phys. Chem. *72*, 4651 (1968)
114) Gopal, R., Rizvi, S. A.: Z. Phys. Chem. (NF) *81*, 330 (1972)
115) Dawson, L. R., Wilhoit, E. D., Holmes, R. R., Sears, P. G.: J. Amer. Chem. Soc. *79*, 3004 (1957)
116) Singh, J. R.: Z. Phys. Chem. (NF) *91*, 317 (1974)
117) Dawson, L. R., Vaughn, J. W., Lester, G. R., Pruitt, M. E., Sears, P. G.: J. Phys. Chem. *67*, 278 (1963)
118) Dawson, L. R., Berger, J. E., Eckstrom, H. C.: J. Phys. Chem. *65*, 986 (1961)
119) Dawson, L. R., Vaughn, J. W., Pruitt, M. E., Eckstrom, H. C.: J. Phys. Chem. *66*, 2684 (1962)
120) Dawson, L. R., Wharton, W. W.: J. Electrochem. Soc. *107*, 710 (1960)
121) Singh, R. D.: Z. Phys. Chem. (NF) *90*, 34 (1974)
122) Edmonson, K. A.: Ph. D. Dissertation, University of Kentucky, Lexington, Ky., 1967; calculation using data from Vaughn, J. W.: Ph. D. Dissertation, University of Kentucky, Lexington, Ky., 1959
123) Dawson, L. R., Wilhoit, E. D., Sears, P. G.: J. Amer. Chem. Soc. *78*, 1569 (1956)
124) Singh, R. D.: Bull. Chem. Soc. Japan *46*, 14 (1973)
125) Dawson, L. R., Lester, G. R., Sears, P. G.: J. Amer. Chem. Soc. *80*, 4233 (1958)
126) Kortüm, G., Quabeck, H.: Ber. Bunsenges. Phys. Chem. *72*, 53 (1968)
127) Fialkov, Yu. Ya., Suprunenko, A. A.: Ukr. Khim. Zh. *41*, 1214 (1975)
128) Fuoss, R. M., Onsager, L.: J. Phys. Chem. *61*, 668 (1957)
129) Fuoss, R. M.: J. Phys. Chem. *79*, 525 (1975)
130) Fuoss, R. M.: J. Phys. Chem. *80*, 2091 (1976)
131) Sears, P. G., Stoeckinger, T. M., Lemire, R. J.: unpublished results, University of Kentucky (1976)
132) Behret, H., Schmithals, F., Barthel, J.: Z. Phys. Chem. (NF) *96*, 73 (1975)
133) Rastogi, P. P.: Z. Phys. Chem. (NF) *85*, 1 (1973)
134) Rastogi, P. P.: Can. J. Chem. *49*, 2004 (1971)
135) Gopal, R., Bhatnagar, O. N.: J. Phys. Chem. *69*, 2382 (1965)
136) Longsworth, L. G.: J. Amer. Chem. Soc. *54*, 2741 (1932)
137) Notley, J. M., Spiro, M.: J. Phys. Chem. *70*, 1502 (1966)
138) Krumgal'z, B. S.: Russ. J. Phys. Chem. *47*, 528 (1973)
139) Robinson, R. A., Stokes, R. H.: Electrolyte Solutions, 2nd edition revised, p. 465. London: Butterworths (1965)
140) Knecht, L. A.: Ph. D. Dissertation, University of Minnesota, Minneapolis, Minn., 1959
141) Pucci, A. E., Vedel, J., Trémillon, B.: J. Electroanal. Chem. *22*, 253 (1969)
142) Dawson, L. R., Sheridan, R. C., Eckstrom, H. C.: J. Phys. Chem. *65*, 1829 (1961)
143) Sibille, S., Rieul, M., Périchon, J.: Experientia Suppl. *18*, 591 (1971)
144) Gosselet, M., Sibille, S., Périchon, J.: Bull. Soc. Chim. France 249 (1975)
145) Payne, R.: J. Phys. Chem. *73*, 3598 (1969)
146) Sibille, S., Périchon, J.: C. R. Acad. Sci. Ser. C *273*, 859 (1971)
147) Sellers, D. E., Leonard, G. W., Jr.: Anal. Chem. *33*, 334 (1961)
148) Agurto-Cid, L. B., Machtinger, M.: Bull. Soc. Chim. France 1915 (1965)
149) Thompson, D. M.: Ph. D. Dissertation, University of Kentucky, Lexington, Ky., 1972
150) Couch, D. E.: Electrochim. Acta *9*, 327 (1964)
151) Schmidt, P. F., Michel, W.: J. Electrochem. Soc. *104*, 230 (1957)
152) Haas, W.: J. Electrochem. Soc. *109*, 1192 (1962)

153) Duffek, E. F., Mylroie, C., Benjamini, E. A.: J. Electrochem. Soc. *111*, 1042 (1964)

154) Okada, M.: Chubu Kogyo Daigaku Kiyo *10A*, 125 (1974); Chem. Abst. *83*, 122941s (1975)

155) Harrick, N. J.: Ann. N. Y. Acad. Sci. *101*, 928 (1963)

156) Müller, H., Eisen, F. H., Mayer, J. W.: J. Electrochem. Soc. *122*, 651 (1975)

157) Brennan, L. C., Gri, N. J.: J. Electrochem. Soc. *117*, 391 (1970)

158) Panzer, R. E.: U. S. Patent 3,117,032 (1964)

159) Dawson, L. R., Berger, J. E., Vaughn, J. W., Eckstrom, H. C.: J. Phys. Chem. *67*, 281 (1963)

160) Tanner, D. W., Berry, G. C.: J. Polym. Sci. Polym. Phys. Ed. *12*, 941 (1974)

161) Walker, E. E.: J. Appl. Chem. (London) *2*, 470 (1952)

162) Bonner, O. D., Bunzl, K. W., Woolsey, G. B.: J. Phys. Chem. *70*, 778 (1966)

163) Chandra, D., Gopal, R.: J. Indian Chem. Soc. *45*, 351 (1968)

164) Koepp, H. M., Wendt, H., Strehlow, H.: Z. Elektrochem. *64*, 483 (1960)

165) Alfenaar, M.: J. Phys. Chem. *79*, 2200 (1975)

166) Alexander, R., Parker, A. J., Sharp, J. H., Waghorne, W. E.: J. Amer. Chem. Soc. *94*, 1148 (1972)

167) Diggle, J. W., Parker, A. J.: Electrochim. Acta *18*, 975 (1973)

168) De Visser, C., Somsen, G.: J. Chem. Thermodynam. *4*, 313 (1972)

169) De Visser, C., Somsen, G.: J. Chem. Thermodynam. *5*, 147 (1973)

170) De Visser, C., Somsen, G.: J. Chem. Soc. Faraday Trans. I *69*, 1440 (1973)

171) Tomus, E. J.: Stud. Cercet. Chim. *18*, 723 (1970)

172) Bonner, O. D., Woolsey, G. B.: Tetrahedron *24*, 3625 (1968)

173) Bonner, O. D., Kim, S. J., Torres, A. L.: J. Phys. Chem. *73*, 1968 (1969)

174) Kreis, R. W., Wood, R. H.: J. Phys. Chem. *75*, 2319 (1971)

175) reference 139, pp. 483–485

176) Lindenbaum, S., Boyd, G. E.: J. Phys. Chem. *68*, 911 (1964)

177) Wen, W. Y., Saito, S., Lee, C. M.: J. Phys. Chem. *70*, 1244 (1966)

178) Robinson, R. A.: J. Phys. Chem. *73*, 3165 (1969)

179) Falcone, J. S., Jr., Wood, R. H.: J. Solution Chem. *3*, 215 (1974)

180) Friedman, H. L.: Ionic Solution Theory, Monographs in Statistical Physics 3. New York: Interscience (1962)

181) Wu, Y. C., Friedman, H. L.: J. Phys. Chem. *70*, 166 (1966)

182) Lindenbaum, S.: J. Phys. Chem. *70*, 814 (1966)

183) Falcone, J. S., Jr., Wood, R. H.: J. Solution Chem. *3*, 239 (1974)

184) Gopal, R., Siddiqi, M. A.: J. Phys. Chem. *73*, 3390 (1969)

185) Gopal, R., Siddiqi, M. A., Singh, K.: Z. Phys. Chem. (NF) *75*, 7 (1971)

186) Gopal, R., Singh, K., Siddiqi, M. A.: J. Indian Chem. Soc. *47*, 504 (1970)

187) Franks, F., Smith, H. T.: Trans. Faraday Soc. *63*, 2586 (1967)

188) Singh, K., Agarwal, D. K., Kumar, R.: J. Indian Chem. Soc. *52*, 304 (1975)

189) Redlich, O., Meyer, D. M.: Chem. Rev. *64*, 221 (1964)

190) Millero, F. J.: J. Phys. Chem. *72*, 3209 (1968)

191) Smiley, H. M.: J. Chem. Eng. Data *15*, 413 (1970)

192) Frost, R., Bittrich, H. J.: Wiss. Z. Tech. Hochsch. Chem. "Carl Schorlemmer" *16*, 18 (1974)

193) Christensen, J. J., Izatt, R. M., Hansen, L. D.: J. Amer. Chem. Soc. *89*, 213 (1967)

194) Weast, R. C. (editor): Handbook of Chemistry and Physics, 47th edition, pp. D85–D87, Cleveland, Ohio: The Chemical Rubber Co. (1966)

195) Casteel, J. F., Amis, E. S.: J. Phys. Chem. *77*, 688 (1973)

196) Vaughn, J. W.: Ph. D. Dissertation, University of Kentucky, Lexington, Ky., 1959

197) Reynaud, R.: C. R. Acad. Sci. Ser. C *266*, 1623 (1968)

198) Carr, W., Shutt, W. J.: Trans. Faraday Soc. *35*, 579 (1939)

199) Halle, J. C., Harivel, R., Gaboriaud, R.: Can. J. Chem. *52*, 1774 (1974)

200) Verhoek, F. H.: J. Amer. Chem. Soc. *58*, 2577 (1936)

201) Guiot, S., Trémillon, B.: J. Electroanal. Chem. *18*, 261 (1968)

202) reference 139, pp. 303, 304

203) Jones, G., Dole, M.: J. Amer. Chem. Soc. *51*, 2950 (1929)

204) Sears, P. G.: Ph. D. Dissertation, University of Kentucky, Lexington, Ky., 1953

205) Falcone, J. S., Jr.: Ph. D. Dissertation, University of Delaware, Newark, Delaware, 1972

206) Singh, R. D., Husain, M. M.: Z. Phys. Chem. (NF) 94, 193 (1975)
207) reference 139, p. 128
208) Glavaš, M.: Therm. Anal. Proc. 3rd Intl. Conf. 1971 2, 341 (1972)
209) Wayland, B. B., Schramm, R. F.: Inorg. Chem. 8, 971 (1969)
210) Haas, D. J.: Nature 201, 64 (1964)
211) Rao, C. N. R.: J. Mol. Struct. 19, 493 (1973)
212) Baron, M. H., De Lozé, C.: J. Chim. Phys. 69, 1084 (1972)
213) Bellô, J., Haas, D., Bello, H. R.: Biochem. 5, 2539 (1966)
214) Toman, K., Honzl, J., Ječný, J.: Acta Cryst. 18, 673 (1965)
215) Rao, C. N. R., Randhawa, H. S., Reddy, N. V. R., Chakravorty, D.: Spectrochim. Acta 31A, 1283 (1975)
216) Martinette, M., Mizushima, S., Quagliano, J. V.: Spectrochim. Acta 15, 77 (1959)
217) Gosavi, R. K., Rao, C. N. R.: J. Inorg. Nucl. Chem. 29, 1937 (1967)
218) Doskočilová, D., Schneider, B.: Coll. Czech. Chem. Commun. 27, 2605 (1962)
219) Matsui, Y., Date, Y.: Bull. Chem. Soc. Japan 43, 2828 (1970)
220) Sibille, S., Yu, L. T., Périchon, J., Buvet, R.: C. R. Acad. Sci. Ser. C 265, 1380 (1967)
221) Gopal, R., Singh, J. R.: J. Indian Chem. Soc. 49, 667 (1972)
222) Davies, C. W.: Ion Association, Chapter 11, London: Butterworths (1962)
223) Gopal, R., Singh, J. R.: J. Phys. Chem. 77, 554 (1973)
224) Gopal, R., Singh, J. R.: Kolloid Z. Z. Polym. 239, 669 (1970)
225) Wilhoit, E. D.: Ph. D. Dissertation, University of Kentucky, Lexington, Ky., 1956
226) Lester, G. R., Gover, T. A., Sears, P. G.: J. Phys. Chem. 60, 1076 (1956)
227) Prue, J. E., Sherrington, P. J.: Trans. Faraday Soc. 57, 1795 (1961)
228) Reynolds, W. L., Weiss, R. H.: J. Amer. Chem. Soc. 81, 1790 (1959)
229) Drago, R. S., Hart, D. M., Carlson, R. L.: J. Amer. Chem. Soc. 87, 1900 (1965)
230) Balyatinskaya, L. N.: Russ. J. Inorg. Chem. 20, 1773 (1975)
231) Gopal, R., Bhatnagar, O. N.: J. Indian Chem. Soc. 44, 1082 (1967)
232) Rees, E. D., Singer, S. J.: Arch. Biochem. Biophys. 63, 144 (1956)
233) Danzer, L. A., Ades, H., Rees, E. D.: Biochem. Biophys. Acta 386, 26 (1975)
234) Franzen, J. S., Harry, J. B., Bobik, C.: Biopolymers 5, 193 (1967)
235) Harry, J. B., Franzen, J. S.: Biopolymers 8, 433 (1969)
236) Schiffer, D. K., Holtzer, A.: Biopolymers 13, 853 (1974)
237) Eckstrom, H. C., Oei, D. G., Dawson, L. R.: J. Phys. Chem. 67, 2804 (1963)
238) Saitô, S., Yabuki, T., Moriwake, T., Okamoto, K.: Bull. Chem. Soc. Japan 46, 1795 (1973)
239) Horclois, R. J., Metivier, J.: U. S. Patent 2,708,670 (1955)
240) Campbell, T. W.: J. Polym. Sci. 28, 87 (1958)
241) Segaller, D.: J. Chem. Soc. 105, 106 (1914)
242) Hall, H. K., Jr.: J. Amer. Chem. Soc. 78, 2717 (1956)
243) Yoneda, S., Morishima, I., Fukui, K., Yoshida, Z.: Kogyo Kagaku Zasshi 68, 1074 (1965)
244) Smiley, H. M.: Ph. D. Dissertation, University of Kentucky, Lexington, Ky., 1960; Dissert. Abst. 26, 4273 (1966)
245) Verhoek, F. H.: J. Amer. Chem. Soc. 56, 571 (1934)
246) Wolford, R. K.: Ph. D. Dissertation, University of Kentucky, Lexington, Ky., 1958; Dissert. Abst. 26, 5071 (1966)
247) Mayhew, R. T., Amis, E. S.: J. Phys. Chem. 79, 862 (1975)
248) Guidoux, R.: Diabetologia 5, 11 (1969)
249) Caujolle, F., Chanh, P. H., Xuong, N. D., Azum-Gelade, M. C.: Arzneim-Forsch. 20, 1242 (1970)
250) Von Kreybig, T., Pruessman, R., Schmidt, W.: Arneim-Forsch. 18, 645 (1968)
251) Peters, G., Guidoux, R., Grassi, L.: Naunyn-Schmiedebergs Arch. Pharmak. Exp. Pathol. 255, 58 (1966)
252) Sitt, R., Senft, G., Losert, W., Barthelheimer, H. K.: Naunyn-Schmiedebergs Arch. Pharmak. Exp. Pathol. 255, 383 (1966)
253) Iynedjian, P. B., Peters, G.: Naunyn-Schmiedebergs Arch. Pharmacol. 280, 209 (1973)
254) Wallon, D., Browaeys, J., Lechat, P.: Sang 31, 871 (1960)
255) Thiersch, J. B.: Inter. Congr. Chemotherapy Proc., 3rd Stuttgart 1963, p. 1741 (1964)

256) Dawson, L. R., Oei, D. G.: J. Colloid Chem. *20*, 282 (1965)
257) Chalkley, L.: U. S. Patent 2,936,235 (1960)
258) Yudelson, J. S. (Eastman Kodak): French Patent 1,463,243 (1966)
259) Kraitr, M., Komers, R., Čuta, F.: J. Chromatogr. *86*, 1 (1973)
260) Kraitr, M., Komers, R., Čuta, F.: Coll. Czech. Chem. Commun. *39*, 1440 (1974)
261) Geipel, H., Rudnick, K., Fischer, K.: German (East) Patent 106,194 (1974)

Received August 1, 1977

EROS

A Computer Program for Generating Sequences of Reactions

Johann Gasteiger and Clemens Jochum

Institut für Organische Chemie der Technischen Universität München, Germany

A computer program has been developed which automatically generates sequences of chemical reactions. As reactions are treated quite formally, all conceivable reactions are obtained irrespective of whether they are already known or are without precedence. This program is especially useful for planning organic syntheses and for generating the products of starting materials. The system is in very active development but the present version can already serve as a tool for the chemist in studying problems of immediate interest. Examples for applications are given to demonstrate the flexibility and performance of the program.

Table of Contents

J. Gasteiger and C. Jochum

1. Introduction

Until recently the application of computers in chemistry was restricted to two fields. First, there were numerical calculations as for the analysis of experimental data and in quantum mechanical computations, and second, computers were used for documentation and information retrieval.

For the past few years problems of immediate chemical interest have been attacked with the aid of computers. The design of organic synthesis, structure-activity correlations, and the elucidation of molecular structure from spectroscopic data are some of the more prominent topics. Here, problems are handled which were thought to be solvable only by the human mind. But it has been recognized that human intelligence can be supported by artificial intelligence, particularly in areas as complex, broad, and expanding as chemistry. Thus, methods of artificial intelligence, used mainly for the simulation of games, suddenly find new, scientifically and economically important fields of application.

Among these the design of syntheses for organic compounds is probably the most interesting and challenging problem. We have a wide spectrum of synthetic problems, depending not only on the classes of compounds to be synthesized but also on the amount of product desired and the accessible reaction conditions. Sometimes only a few grams of a compound have to be synthesized in the laboratory to study physical, chemical, or biological properties. On the other hand, syntheses are done on a huge scale in industrial processes. In spite of the enormous importance of organic synthesis, there are only a few approaches to systematic study of the planning of syntheses. This may be due to the large amount of information to be taken into account and the many alternatives and decision processes involved. Further, in synthetic design the analysis has to be done in the reverse direction of the actual synthetic reaction in the laboratory. For this backward strategy the term retrosynthetic approach seems to have been generally accepted. We have been trained to understand reactions in the direction in which they are proceeding; the retrosynthetic reasoning still has to wait for a general systematization and has not yet entered into teaching.

In planning the synthesis of an organic compound, the structure of the desired target has to be analyzed, methods for the preparation of the skeleton and the functional groups have to be devised, and the relative merits of these reactions have to be estimated. This must be done not only for one reaction but for whole sequences of synthetic reactions. Some of these sequences cannot be evaluated separately but their importance is dependent on the outcome of other sequences. Thus, complex strategic considerations are involved which demand a high degree of insight, the generation and evaluation of intermediates, reactions, and paths as a whole, and the testing of hypotheses and decisions.

The large amount of information to be processed and the requirement for decisions among a host of alternatives suggest the use of computers in synthetic planning. But still quite often the opinion prevails that the inventiveness of the chemist cannot expect any assistance from the computer. This challenge, and the importance of synthetic planning, are responsible for the development of computer programs for the planning of organic syntheses.

Around 1967 three groups[1–3] started to attack this problem in a way not restricted to special classes of molecules. In handling the formal aspects of chemical reactions by computers, one has to decide on representations for molecules and reactions. While the choice of the representation of molecules is more or less a technical problem, provided that no information of chemical interest is disregarded, the selection of a representation for chemical reactions is of central importance and can be decisive for the inventiveness of the program system. All three systems which were conceived in 1967 – LHASA[1], SECS[2], and SYNCHEM[3] – have in common that the synthetic reactions are taken from a collection of known reactions, a library which has to be built up in the computer.

By analyzing an input molecule, substructures are found which point to chapters in the reaction library. There, descriptions of changes in these substructures are contained which will occur in the course of a certain reaction. Implementing these changes leads to synthetic precursors. Thus, the recognition of the substructure of a β-hydroxy-carbonyl group would point to the chapter aldol condensation in the reaction library. There the structural changes of a retro-aldol condensation are contained and result in the two carbonyl compounds as precursors (see Fig. 1). As the reac-

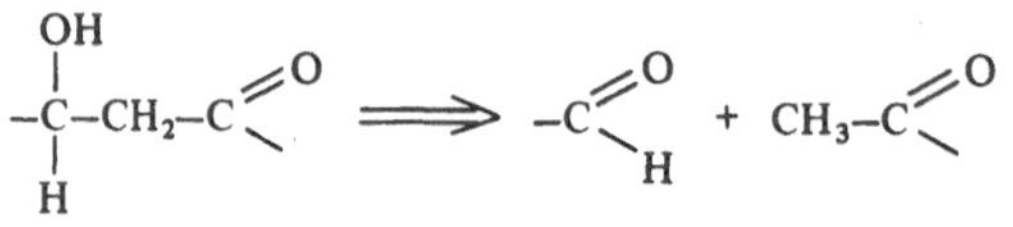

Fig. 1. Aldol condensation, as taken from a reaction library

tion library has to be generated by the program designer, it will always be restricted to his knowledge of chemical reactions; in the ideal case – which will hardly be achievable – it will be limited to all known chemistry at the time of creation of the reaction library. It is inherent in such an approach that only syntheses as combinations of known reactions will be found. Further, as new synthetic reactions are discovered, the library has steadily to be enlarged leading to bigger core requirements and longer search times.

When our group started in 1972 to develop a computer program for the design of syntheses, we based it on a mathematical model of constitutional chemistry[4, 5]. In this model molecules are represented by BE-matrices (see Section 2.1.), reactions are taken as transformations of BE-matrices of isomeric ensembles of molecules by R-matrices (see Section 2.2.). These R-matrices indicate the changes in the distribution of bonds and free electrons occurring in a reaction. The R-matrices can be generated quite formally thereby obtaining all conceivable reactions irrespective of whether a reaction is already known or is completely novel, i.e., without any precedence. Thus, the limitation to known chemistry is removed.

This mathematical model can serve as a basis for a variety of deductive computer programs for the solution of chemical problems[6–8]. From among all the possibilities, synthetic planning was chosen first – for reasons stated before – and the computer program CICLOPS (Computers in Chemistry, Logic Oriented Planning of Syntheses)[9-11)] was developed. Relying on the insights we gained with this prototype program, we designed and implemented our new synthetic planning program EROS (Elaboration of Reactions for Organic Synthesis). EROS is under constant expansion and improvement. It can be applied not only to retrosynthetic analysis as in synthetic design but also to the generation of sequences of forward reactions.

Other groups have entered the field of computer-assisted synthetic planning[12)]. The approaches of Bersohn[13)], of Barone, Chanon and Metzger[14)], and of Blower and Whitlock[15)] all rely on libraries of known reactions to generate synthetic precursors. The computer program developed by Weise[16)], on the other hand, is based on the same mathematical model of chemistry[4, 5)] as our system. However, the structure and implementation of the two systems are entirely different. Yoneda[17)] has independently developed a matrix representation for molecules and reactions quite similar to the one in the mathematical model used by us. This representation serves as a basis for a computer program to study heterogeneous catalysis of gas phase reactions. Hendrickson has developed formal concepts for synthesis design[a)], but implementation of his ideas in a computer program is still lacking.

2. A Mathematical Model of Chemistry and Its Implementation

For generating synthetic pathways by a computer, representations for the intermediates of chemical syntheses (molecules) and for the transformation intermediates of these (reactions) have to be designed. Both molecules and reactions are represented in our model of constitutional chemistry[4, 5)] by matrices. Internally, in the program, more compact forms are used.

2.1. Molecular Structures

The mathematical model chosen as a basis is confined to constitutional chemistry. The formal concepts for the treatment of the stereochemical features of molecules and reactions have been given elsewhere[10)], and a computer program based on these ideas is being developed. For the initial phase of the development of a computer program for the design of syntheses, stereochemistry has temporarily been set aside to concentrate on the more important aspects of constitutional chemistry.

The constitution of a molecule is given by the set of its atoms and the bonds between these atoms. When also taking into account the number of free electrons, double-bookkeeping of all valence electrons in a reaction becomes possible[5)]: the sum of free and bonding electrons has to stay constant. In the present mathematical model[4, 5)] molecules are represented by so-called BE-matrices (bond and electron matrices). The rows and columns of these matrices refer to the atoms of a molecule; the off-diagonal entries give the bond orders between the respective atoms, the diagonal entries show the number of free electrons on an atom. Figure 2 serves as an

---

[a)] Hendrickson, J. B.: Topics Curr. Chem. 62, 49 (1976).

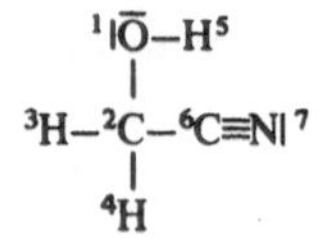

$$^1|\bar{O}-H^5$$
$$|$$
$$^3H-^2C-^6C\equiv N|^7$$
$$|$$
$$^4H$$

1

	1	2	3	4	5	6	7
	O	C	H	H	H	C	N
1 O	4	1	0	0	1	0	0
2 C	1	0	1	1	0	1	0
3 H	0	1	0	0	0	0	0
4 H	0	1	0	0	0	0	0
5 H	1	0	0	0	0	0	0
6 C	0	1	0	0	0	0	3
7 N	0	0	0	0	0	3	2

Fig. 2. BE-matrix of the cyanohydrin of formaldehyde 1 (atoms are numbered arbitrarily)

illustration for a *BE*-matrix. For example, the elements (6, 7) and (7, 6) of the matrix represent the triple bond of the cyano group, element (1, 1) shows the four free electrons on the oxygen atom.

A *BE*-matrix has a number of interesting mathematical properties that directly refer to chemical facts. As these can be found in the literature[4–6], they will not be discussed in detail here. Suffice to say that one can easily find the total number of bonds of an atom and the charge on an atom. Further, it can be checked whether the octet rule is obeyed. *BE*-matrices can be constructed not only for single molecules but also completely analogously for a collection of several molecules, *i.e.*, for an ensemble of molecules.

Figure 3 gives the *BE*-matrix of the ensemble of formaldehyde and hydrocyanic acid.

The basis for the formulation of *BE*-matrices for ensembles of molecules is an extension of the concept of isomerism to ensembles of molecules: The cyanohydrin of formaldehyde *1* is isomeric with the ensemble of its components formaldehyde *2* and hydrocyanic acid *3*. Both *1* and *2* + *3* have the same set of atoms and the same number of valence electrons.

Many entries in a *BE*-matrix are zero. Therefore, for the internal representation of molecules in the program a more compact form was chosen. The constitution of molecules and ensemble of molecules is given by doubly indexed lists of bonds. The free electrons on the atoms are kept in a separate vector, as are the atomic numbers (see Fig. 4).

In the bond list the first two rows give the number of the two atoms participating in a bond, the third row contains the bond order. For example, the first column (1 2 1) shows that atom number 1 (the oxygen atom) is bonded to atom number 2 (a carbon atom) by a single bond. The bond list could be further compressed as each bond shows up twice. But this small waste in storage space is balanced by faster

$$^3\text{H}\diagdown\!\!\!\!_{} ^2\text{C}{=}\bar{\text{O}}^1 \quad + \quad {}^5\text{H}{-}{}^6\text{C}{\equiv}\text{N}|^7$$
$$^4\text{H}\diagup$$

$$2 \qquad\qquad\qquad 3$$

| | | 1 | 2 | 3 | 4 | 5 | 6 | 7 |
		O	C	H	H	H	C	N
1	O	4	2	0	0	0	0	0
2	C	2	0	1	1	0	0	0
3	H	0	1	0	0	0	0	0
4	H	0	1	0	0	0	0	0
5	H	0	0	0	0	0	1	0
6	C	0	0	0	0	1	0	3
7	N	0	0	0	0	0	3	2

Fig. 3. BE-matrix for the ensemble of formaldehyde **2** and hydrocyanic acid **3**; (atom numbers as in Fig. 2)

Atom Number	1	2	3	4	5	6	7
Atomic Number	8	6	1	1	1	6	7
Free Electrons	4	0	0	0	0	0	2

Bond List	1. Atom	1	1	2	2	2	2	3	4	5	6	6	7
	2. Atom	2	5	1	3	4	6	2	2	1	2	7	6
	Bond Order	1	1	1	1	1	1	1	1	1	1	3	3

Fig. 4. Internal representation of the cyanohydrin of formaldehyde **1**; (atom numbers as in Fig. 2)

access to row sums which are necessary for determining all bonds or neighbors of an atom.

2.2. Chemical Reactions

The *BE*-matrices of Figs. 2 and 3 are representations for the molecules of the starting and end points of a chemical reaction, the formation of the cyanohydrin *1* of formaldehyde from its components *2* and *3*. Taking the difference between these two matrices — from each element of the first matrix the corresponding element of the second matrix has to be subtracted — one obtains another matrix (see Fig. 5). This matrix is a representation of the reaction itself, and is therefore called a reaction matrix or *R*-matrix. Rearranging the matrix equation of Fig. 5 gives Fig. 6.

This equation (Fig. 6) gives a better insight into the nature of an *R*-matrix. The entries of an *R*-matrix indicate which bonds are broken (negative elements) and

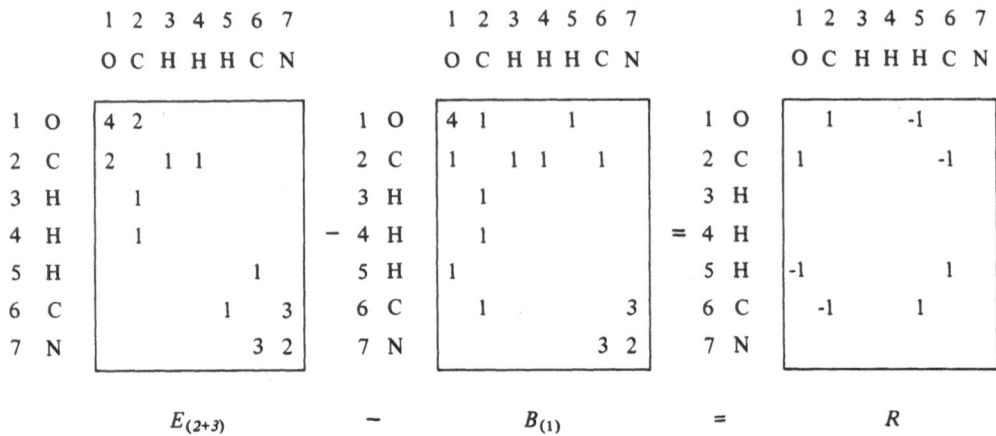

Fig. 5. Matrix equation, resulting in an R-matrix; (for clarity, only the nonzero elements are given)

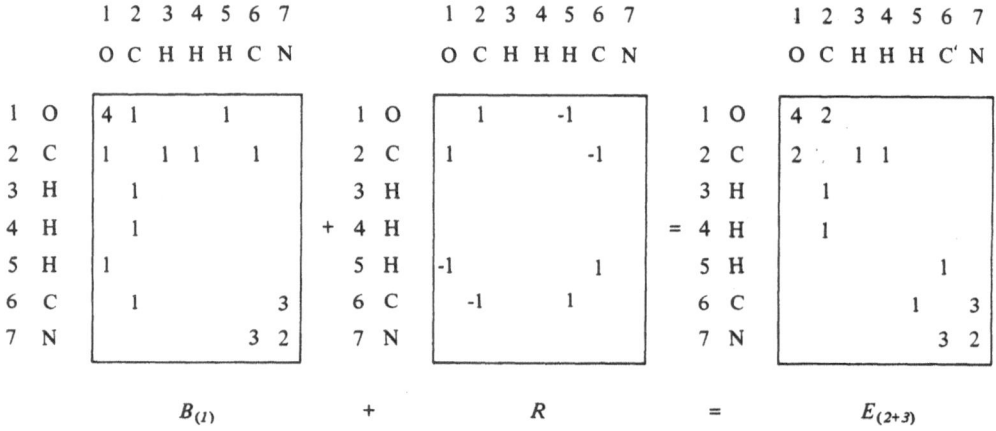

Fig. 6. Matrix equation for the decomposition of the cyanohydrin 1 into its components (only nonzero entries given)

which are made (positive elements) in a reaction. Further, rearrangements in the distribution of the free electrons are given analogously in the diagonal entries. In the present example (Fig. 6) the bonds between oxygen-1 and hydrogen-5 and between carbon atoms 2 and 6 are broken while bonds between oxygen-1 and carbon-2 and between H-5 and C-6 are made. R-matrices have a number of interesting mathematical properties which can be found in the literature[4−6]. For example, the sum of the entries of an R-matrix must be zero. This reflects the fact that in a reaction no electrons are lost or gained.

As we have seen, R-matrices indicate how bonds and electrons are redistributed in a reaction. These electron and bond shifting schemes can be formally generated and classified. For closed-shell chemistry, 38 R-matrices breaking one, two, or three bonds have been found[9, 10]. These R-matrices are of different chemical importance; some represent mechanistic steps, others comprise powerful synthetic reactions. A study on the relative importance of the R-categories based upon the number of rearrangement reactions found in the literature for the various schemes has been published recently[18]. As of now EROS contains three reaction categories. It is estimated that thereby about 90% of all synthetically interesting reactions are covered, included those already known and those to be discovered in the future. In Fig. 7 a very important reaction category is shown. The majority of synthetic reactions falls

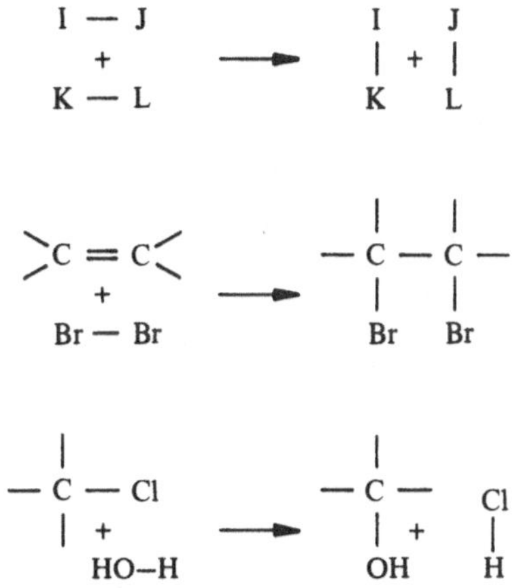

$$
\begin{array}{ccc}
\mathrm{I} - \mathrm{J} & & \mathrm{I} \quad \mathrm{J} \\
+ & \longrightarrow & | \; + \; | \\
\mathrm{K} - \mathrm{L} & & \mathrm{K} \quad \mathrm{L}
\end{array}
$$

$$
\begin{array}{ccc}
{>}\mathrm{C} = \mathrm{C}{<} & & -\overset{|}{\underset{|}{\mathrm{C}}} - \overset{|}{\underset{|}{\mathrm{C}}} - \\
+ & \longrightarrow & \\
\mathrm{Br} - \mathrm{Br} & & \mathrm{Br} \quad \mathrm{Br}
\end{array}
$$

$$
\begin{array}{ccc}
-\overset{|}{\mathrm{C}} - \mathrm{Cl} & & -\overset{|}{\mathrm{C}} - \quad \mathrm{Cl} \\
| \; + & \longrightarrow & | \; + \; | \\
\mathrm{HO}-\mathrm{H} & & \mathrm{OH} \quad \mathrm{H}
\end{array}
$$

Fig. 7. Reaction category breaking two bonds and making two bonds; examples of addition and substitution reactions

into this category; examples are additions to multiple bonds, substitution reactions, and eliminations. The formation (and decomposition) of the cyanohydrin *1* also belongs to this category (see Fig. 6).

Internally in the program, these reaction categories are not handled as full matrices. Rather, only the nonzero elements are used. The negative entries, corresponding to the breaking of bonds, are directly applied to the list of bonds. The positive elements, which show the new bonds, are put into a separate list of bonds made which is combined with the modified list thus yielding the list of the bonds of the products. Changes in the distribution of free electrons are directly made on the corresponding vector.

The internal generation of a reaction shall be illustrated in Fig. 8 with the decomposition of the cyanohydrin *1*. The reaction category of Fig. 7 will be applied; only those parts of the molecule and of the list of bonds are shown which are directly involved in the reaction process. On the right-hand side of Fig. 8 the general scheme and the elements of the reaction category are indicated; the left part shows the specific example chosen. After selecting breakable bonds (see Chapter 3), O^1-H^5 and C^2-C^6, these bonds are rearranged according to the scheme on the right (and in Fig. 7), *i.e.*, these two bonds are broken and two new ones are made. The list of bonds thus obtained is already the one of the ensemble of formaldehyde *2* and hydrocyanic acid *3*.

These reaction products (or more precisely: products of the retro-reaction) were obtained by applying a formal scheme. In relocating the bonds no information on the specific reaction at hand was needed. In generating the reaction products it was

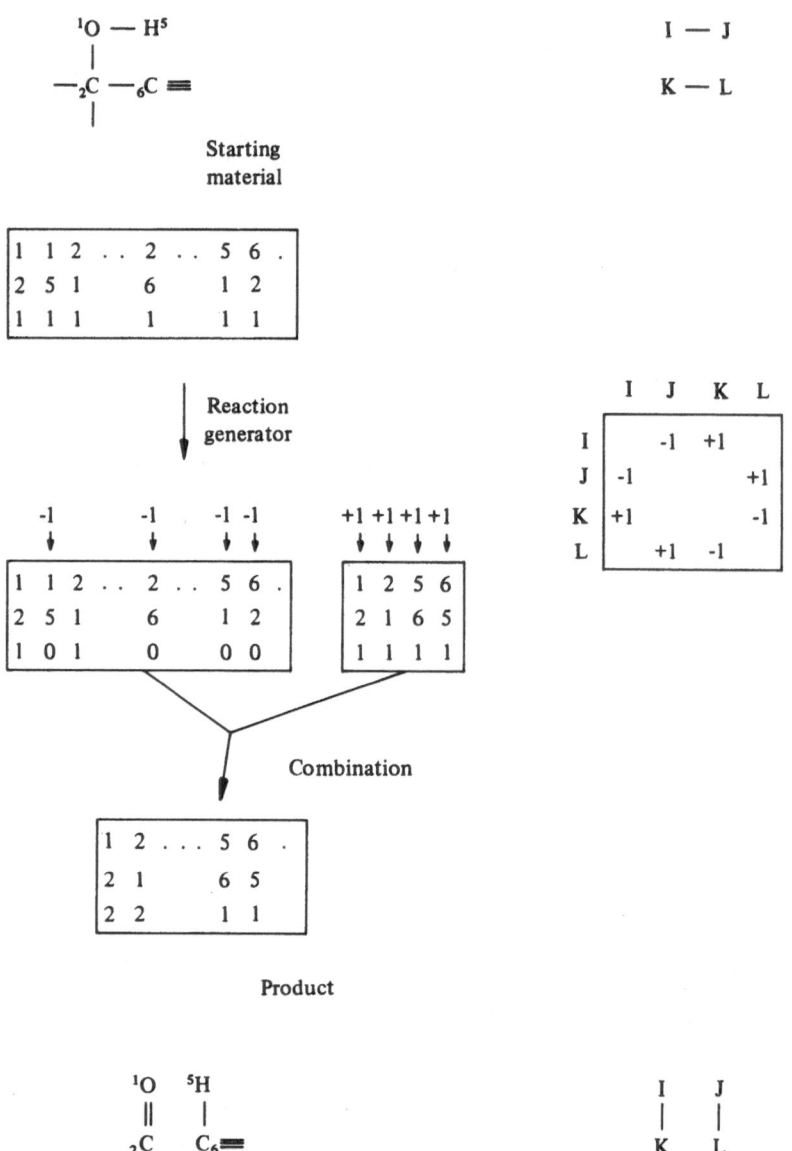

Fig. 8. Generation of reactions

completely irrelevant whether this reaction was already known or entirely novel. Thus, our approach does not suffer from being restricted to known chemistry as is the case with programs based on reaction libraries.

Applying the general reaction schemes contained in our program system onto the bonds of molecules allows the generation of all conceivable reactions or retro-reactions of these molecules. The task is then to select the chemically feasible reactions from the set of the mathematically possible ones. Here lie the main efforts in the development of our program. Chemical knowledge is condensed into general rules which permit evaluation or selection of individual reactions. Among the various selection criteria the definition of the reaction site is one of the most interesting fields of research.

3. Reaction Site

The reaction site is defined through breakable bonds. The general reaction schemes are only applied to bonds which are considered to be breakable. The concept of breakable bonds is an approach to dealing with chemical reactivity.

The organic chemist has learned to look at the reactivity of molecules in terms of the reactivity of its functional groups. Most textbooks in organic chemistry are organized according to functional groups, and reactions are taken as interconversions of functional groups. Learning the reactivities of functional groups has given most chemists the false impression that they understand the reactivity of molecules. But, in fact, it has often prevented the seeking of general guidelines and a deeper under-standing of chemical reactivity. In our system, functional groups are not treated explicitly but their reactivity is rather seen in terms of electronic effects of individ-ual bonds and their interactions.

One of our early approaches might serve to illustrate the routes we are follow-ing in coping with this problem. As "breakable bonds" we considered: multiple bonds, bonds to heteroatoms, as well as bonds which are one or two neighbors away from such bonds. The reason for choosing these bonds was that multiple bonds and heteroatoms are responsible for perturbations in a uniform electronic distribution of a molecule and thus offer sites for electrophilic, nulceophilic, or radical attack. Admittedly, this is a very crude scheme; it serves only as a starting point for con-tinuing research. More refined schemes are being developed.

To further explain the concept of breakable bonds let us return to the example in the Introduction (see Fig. 1). For the substructural unit of a β-hydroxy-carbonyl group EROS would define — among others — the OH-bond and the CC-bond in β-position to the carbonyl group as breakable. Now, applying the reaction scheme we have already encountered several times — breaking two bonds and making two (see Fig. 7) — leads to the precursors of the aldol condensation (see Fig. 9). Again, it should be stressed that these reaction products were obtained here without mak-ing use of any knowledge of the aldol condensation. The search for breakable bonds and the application of the reaction scheme lead to the same net result.

At present, various options for defining breakable bonds are available on input. There is a choice of the kind of bond — multiple bonds, bonds to heteroatoms,

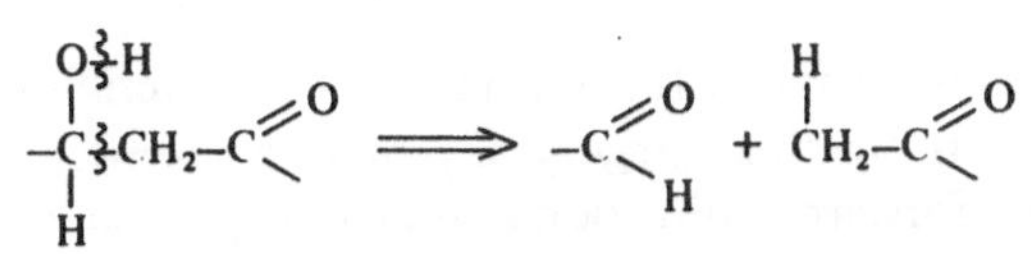

Fig. 9. Retro-aldol condensation; breakable bonds as indicated

and/or bonds adjacent to these — which will be treated as breakable. Further, bonds in the input molecules can directly be taken as being breakable to meet special problem requirements and reduce the amount of output to be searched. This might seem as an undue restriction imposed on the run of the program. But EROS should not be used as a black box; it is rather a device given into the hands of the chemist to aid in the solution of his problems. The range of problems it can be applied to is enormous — from synthesis design to the generation of reaction sequences. The more specific a question is defined the lesser the output that has to be scanned to find the best solutions. Furthermore, once the choices have been made the program runs without interference by the user to produce multistep reactions. Thus, the prejudices of the chemist user resulting from his personal training and his necessarily limited knowledge of chemistry are eliminated. This is not the case with interactive programs where the chemist is required to interfere in the selection process to direct the program to realistic solutions of synthetic problems.

After the selection of breakable bonds the program determines which bond- and electron rearrangements are possible on distinct atoms. This is achieved by checking a table of allowed valence configurations, *i. e.,* information on permitted combinations of numbers of free electrons and numbers of bonds for each atom type. Thus, an atom having achieved its maximum number of bonds will not be submitted to a reaction scheme which would add another bond to that atom. Different tables can be used according to the problem at hand. A restrictive table allows only species without charges; this is sufficient for most problems in synthesis design. A second table is more detailed, permitting also charge configurations like carbanions and carbonium ions, important for studies of reaction mechanisms.

Both the selection of breakable bonds and the consideration of valence changes can be carried out before actually applying the reaction schemes. This has the effect that mathematically possible but chemically not feasible reactions are eliminated before they are generated. This ideal case in the selection process cannot be achieved with every criterion of evaluation. With some criteria the reaction has first to be generated, and the products are discarded only after subsequent evaluation (see Chapter 5).

4. Reaction Partners

In our system, reactions are considered as isomerizations of ensembles of molecules; both sides of the reaction equation contain the same set of atoms and the same number of valence electrons. Then, for generating reactions, all compounds which can conceivably take part in these reactions have to be known. Generally there is not only one compound reacting — exceptions are rearrangements or decompositions — but there are several reaction partners.

As EROS can be applied to two entirely different types of problems, the definition of reaction partners has two aspects. In the first category of problems, reactions are accepted as proceeding in the same — forward — direction as they were generated. Then, from starting compounds, sequences of reaction products are obtained. In these cases the definition of the reaction partners usually poses no problems: These are the compounds that are brought together and whose products one wants to know. In synthetic planning, on the other hand, one is working retrosynthetically. The compounds generated are actually the starting materials needed to synthesize the input molecule. Here, now, the compounds with which the target has to be brought into "reaction" are not immediately obvious. In this case, the "reaction partners" are in fact compounds which are generated together with the target molecule in the synthetic reaction considered. It seems misleading to call these compounds by-products because one usually associates with this name compounds which are formed in side reactions. Here, these compounds which are generated in the very same reaction event as the target or synthetic intermediate are referred to as "synthesis partners".

An example should serve to clarify the situation: Let us assume that the problem is to find a synthesis for ethyl acetate. Applying the reaction scheme of Fig. 7 on the input structure with the C–H bond (in β-position to the carbonyl group!) and the C–O bond as breakable leads to ketene and ethanol as synthetic precursors (Fig. 10). Of course, this is an approach to the synthesis of ethyl acetate. But just

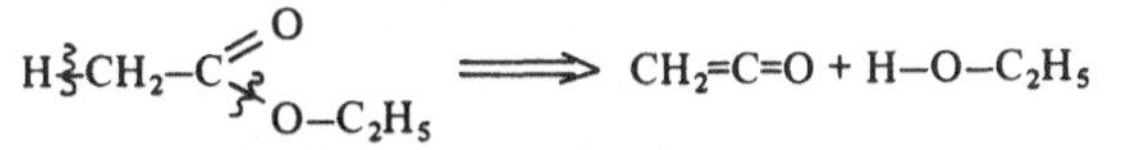

$$H_3^3CH_2-C\!\!\diagup_{O-C_2H_5}^{O} \quad \Longrightarrow \quad CH_2=C=O + H-O-C_2H_5$$

Fig. 10. Synthesis of ethyl acetate, without synthesis partners

working on the structure of ethyl acetate alone, one would never find the common procedure for making this ester, *i.e.*, from acetic acid and ethanol. To find that synthesis one has to augment the target structure with the elements of water. Then, with the same reaction scheme as previously used, acetic acid and ethanol are generated as synthetic precursors (Fig. 11).

$$CH_3-C\!\!\diagup_{O-C_2H_5}^{O} \quad + \; H_3^3OH \quad \Longrightarrow \quad CH_3-C\!\!\diagup_{OH}^{O} \quad + \; H-O-C_2H_5$$

Fig. 11. Synthesis of ethyl acetate with water as synthesis partner

Thus, in the second category of problems, one is faced with the task of finding the appropriate synthesis partners with which the target and the synthetic intermediates are to be augmented to obtain an ensemble of molecules which can be transformed to an isomeric ensemble of promising precursors. This task might seem to be prone to complete arbitrariness. But scanning through collections of synthetic reactions, one meets the same synthesis partners again and again. One finds water, carbon dioxide, nitrogen, hydrogen chloride, sodium chloride, etc.; all small molecules which are generated in the same reaction step as the synthesis products. Quite frequently, these molecules are even dropped when writing reaction equations with the tacit assumption that they are too common to be of importance. But as is seen here they are absolutely necessary to obtain all important syntheses. These molecules comprise a limited set of small and common molecules. EROS contains such a standard set, and the target is augmented with these synthesis partners to form an ensemble upon which the reaction schemes are applied. Further, for special problems, other synthesis partners can be input jointly with the target molecule, providing convenience and flexibility.

5. Selection and Evaluation

5.1. General Considerations

As our reaction schemes generate all formally conceivable reactions, the selection of the appropriate and actually occurring reactions from these is of central importance in our approach. Most easily this could be achieved by taking the human user as an interceptor and selector. Then, the chemist user and his enormous knowledge on chemical reactions, compiled in many years of hard training, could be made directly available to the program thereby vastly improving its performance. But the training of the individual user and his knowledge of chemistry is necessarily limited as is even the present overall insight into chemistry. This renders the performance of the program dependent on the individual user and limits the quality, in the ideal case to the chemistry known at present, and in practise to a much smaller part of it.

Completely novel reactions which will be generated by our approach are in danger of being discarded by the chemist user often just because of their very nature of being novel, *e.g.*, having no examples in the literature. The user unfamiliar with such reactions is inclined to reject them, especially if he is given only a short period of time while communicating with the computer. In short, interactive programs actively including the user in the decision process have merits of their own — improving the performance of incomplete programs particularly in the development phase, and reducing psychological barriers in the man-machine interaction. But our aim is more ambitious: to develop a program without interference by the user, a program with genuine artificial intelligence capabilities.

Requiring the program to automatically make the selection of the chemically feasible reactions from the mathematically possible ones demands that chemical reality be exactly modeled in the computer. Naturally, this goal can be reached only through intensive research. Many different models have to be designed and tested, and chemical heuristics must be utilized. Preferentially, recourse will be taken to physical organic chemistry, and well-established concepts and relations will be employed such as electronegativity, kinetic parameters, σ-constants, steric crowding, etc. The early stages of the selection process have already been met in Chapter 4: First, only molecules containing atoms in allowed valence configurations are generated. Then, the definition of breakable bonds leads to further restrictions. The important estimation of thermodynamic data as exemplified by reaction enthalpies will be dealt with in the next paragraph in more detail.

The various evaluation and selection criteria are either of a binary nature, *i.e.*, they allow yes-no decisions leading to immediate acceptance or rejection of a reaction, or they allow the assignment of a numerical value to a reaction which reflects its quality. In combination with upper or lower limits this numerical value may also serve to accept or reject a reaction. These quality factors have to be carried over several levels of the synthesis tree (see Chapter 6).

Because our approach generates all conceivable reactions, the best solutions to a problem must be contained among them. Selection and evaluation must then reduce the output just to these best solutions. Incomplete program development will have incomplete selection. But as long as this selection is not based on wrong assumptions the best solutions will be retained, accompanied by less favorable ones as the scope for making decisions is still broader. Thus, each version of EROS already has its merits, presenting results which contain the best solutions to a problem. Through the stages of program development, more and more of the less favorable solutions will be discarded, zeroing in on the optimum answers.

Variable degrees of sophistication even seem desirable: There are problems in which the set of all solutions is asked for, *e.g.*, generating all valence tautomers with the empirical formula $(CH)_{10}$. Further, planning a synthesis for an industrial compound will necessarily differ remarkably from planning a synthesis which will be done in the laboratory on a small scale. Quite different reaction conditions have to be taken into account and economic considerations will have different emphasis. Therefore, a flexible system is being developed which allows the user to define starting conditions and variable degrees of restraints to tailor the run of the program to the special problem and needs.

5.2. Reaction Enthalpies

The basic approach in implementing physical organic concepts will be illustrated in this paragraph. One of the most important factors in determining the course of reactions is the thermodynamic aspect of the process, in particular, the enthalpy of a reaction. Therefore, we have developed a program which estimates the enthalpy of each reaction generated[19]. In principle, the enthalpy of a reaction could be obtained by performing quantum mechanical calculations on the starting materials and products of a chemical reaction. Such an approach would be extremely expensive (in terms of computer time), applicable only to rather small model systems, and the numerical precision necessary — 1 to 3 kcal/mole can be decisive for the course of a reaction — could be achieved with reliability only in very rare cases. This makes the quantum mechanical approach at present prohibitive.

As the entirely theoretical calculation of reaction energies is not feasible, an empirical method was developed, making use of thermochemical data. In principle, the reaction enthalpy could be obtained as the difference in the heats of formation of starting materials and products. But only for a small fraction of organic compounds have the heats of formation been determined. Such an approach is possible only in special cases. Rather, experimental data were taken and analyzed on the basis of additivity schemes for the estimation of heats of formation, to obtain parameters which are applicable to whole classes of compounds.

As the concept of localized bonds is a quite reasonable picture for the state of a molecule, the heats of formation can be broken down to bond contributions. For most chemical applications to accuracy thus obtained is not satisfactory because the energy of a bond is dependent on its immediate environment. For example, the dissociation energy of a CH-bond varies from 129 kcal/mole in hydrocyanic acid to 85 kcal/mole in toluene. Over the years a variety of schemes has been proposed for the estimation of heats of formation which attain a higher accuracy than simple bond additivity schemes[20]. The two schemes tested for use in the estimation of reaction energies were the group method of Benson[21] and the scheme proposed by Allen[22].

Allen's scheme leads to a program of greater generality and it will be described in more detail. For many classes of molecules, Benson's and Allen's schemes give the same results. Allen's scheme was originally proposed for the estimation of the heats of formation or heats of atomization of saturated hydrocarbons and organic sulfur compounds[22a] and was later extended to other classes of molecules as well[20, 22b, 23].

How can a scheme proposed for the estimation of heats of formation be employed to the calculation of heats of reaction? The idea is that the reaction energy is largely accounted for by changes in the energies of the bonds participating in the reaction. Again, refuge was taken in the model of localized bonds. Our schemes for generating reactions provide direct access to the bonds broken and made in the course of a reaction. These bonds determine the B-terms of Allen's scheme which can be thought of as being largely made up of bond energies and which are the prevailing parameters. Scanning the environment of the bonds involved in a reaction leads to the correction terms of Allen's scheme which may be viewed as bond-bond and atom-atom interaction energies. Taking the difference between the terms for starting materials and products leads to an estimate for the reaction enthalpy. These energies are further corrected for by addition of strain and resonance energies.

The great advantage of the program based on Allen's scheme is that whenever for a certain class of compounds there are not sufficient data available to determine all the parameters necessary, recourse can be taken to a lower level of sophistication. Then only bond energies are taken into account which leads to at least reasonable estimates for reaction enthalpies.

The calculated values for the heats of reaction are, especially for gas phase reactions, of high quality. For those classes of molecules where sufficient primary experimental data are available the differences between experimental and estimated reaction energies are usually between 0.1 to 0.4 kcal/mol. In fact, whereas the program gives an estimate for nearly any reaction — note that 90 different bond types are considered — comparison with experimental values is quite often hampered by the lack of data. Extensive comparisons can be made only with quite simple or small molecules but there is no reason that the method should fail with larger molecules.

The following table (Table 1) shows a sample of products generated by EROS from n-butane, together with experimental[20] and estimated heats of reaction. Usually the differences are below 0.5 kcal/mole. Only for such highly strained molecules as methylcyclopropene and bicyclo[1.1.0]butane are the differences larger; for methyl cyclopropane no experimental value is available. Table 2 shows reaction

Table 1. Isomerizations of n-butane; comparison of estimated and experimental values for the heats of reaction

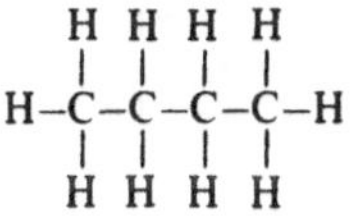

$$H-\underset{\underset{H}{|}}{\overset{\overset{H}{|}}{C}}-\underset{\underset{H}{|}}{\overset{\overset{H}{|}}{C}}-\underset{\underset{H}{|}}{\overset{\overset{H}{|}}{C}}-\underset{\underset{H}{|}}{\overset{\overset{H}{|}}{C}}-H$$

	Heat of reaction (in kcal/mole)	
Products	est.	exp.
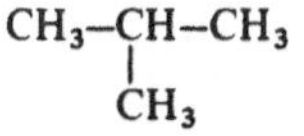 $CH_3-\underset{\underset{CH_3}{\vert}}{CH}-CH_3$	-2.03	-2.05
$CH_3-CH_2-CH=CH_2$ + H_2	+29.92	+30.16
$CH_3-CH=CH-CH_3$ + H_2	+27.15	+27.37
$CH_3-CH=CH_2$ + CH_4	+16.93	+17.35
CH_3-CH_3 + $CH_2=CH_2$	+22.29	+22.57
$CH_2=CH_2$ + $CH_2=CH_2$ + H_2	+54.98	+55.26
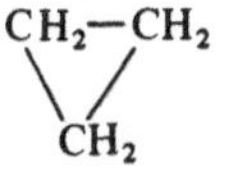 $\underset{CH_2}{CH_2-CH_2}$ + CH_4	+25.03	+25.20
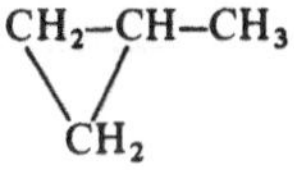 $\underset{CH_2}{CH_2-CH-CH_3}$ + H_2	+35.96	- - - -
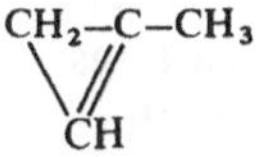 $\underset{CH}{CH_2-C-CH_3}$ + $2H_2$	+87.15	+88.57
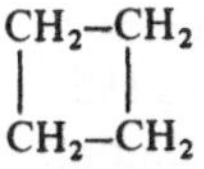 $\underset{CH_2-CH_2}{CH_2-CH_2}$ + H_2	+36.60	+37.14
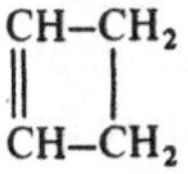 $\underset{CH-CH_2}{CH-CH_2}$ + $2H_2$	+67.35	+67.81
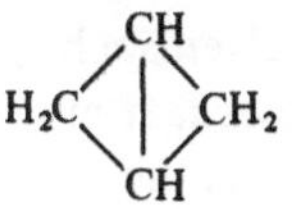 $H_2C \overset{CH}{\underset{CH}{\diamondsuit}} CH_2$ + $2H_2$	+85.14	+82.26

Table 2. Reaction of n-butane with chlorine; comparison of experimental and estimated values for the heats of reaction

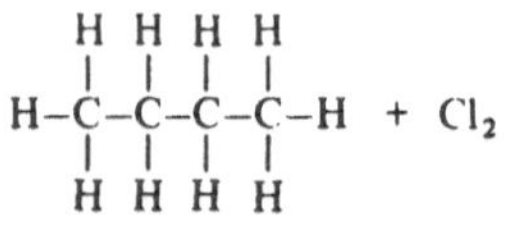

$$
\begin{array}{c}
\text{H H H H} \\
\text{| | | |} \\
\text{H--C--C--C--C--H + Cl}_2 \\
\text{| | | |} \\
\text{H H H H}
\end{array}
$$

Heat of reaction
(in kcal/mole)

Products	est.	exp.
CH_3--CH_2--CH_2--CH_2--Cl + HCl	-27.86	-26.8 (±2.0)
CH_3--CH_2--CH--CH_3 + HCl $\quad$ (Cl)	-31.07	-30.3 (±2.0)
CH_3--CH_2--CH_2--Cl + CH_3Cl	-21.41	-21.19
CH_3--CH_2--CH=CH_2 + 2HCl	-14.17	-13.96
CH_3--CH_2--CH--CH_2--Cl + H_2 $\quad$ (Cl)	-14.83	-- -- --
CH_3--CH_2--Cl + CH_3--CH_2--Cl	-22.04	-23.84
CH_3--CH_2--Cl + CH_2=CH_2 + HCl	-5.58	-5.35
CH_3--CH_3 + Cl--CH_2--CH_2--Cl	-22.04	-20.58

products generated from n-butane and chlorine. Experimental values for chlorine-containing compounds are known with much less accuracy than for hydrocarbons and oxygen containing compounds. For 1-chlorobutane and 2-chlorobutane, the experimental uncertainty is 2.0 kcal/mole. Thus, for these two compounds, the calculated values are well within the experimental error. Extending the scheme to consider also 1.4-interactions would further improve the estimates. The values estimated for the reaction energies can be used for evaluating the reactions. Depending on the kind of problem studied, different approaches have to be taken. When considering only one-step reactions the reaction enthalpy can be taken as the predominant criterion for evaluating different reactions. Provided that thermodynamic product control prevails, the reaction with the lowest enthalpy — the most exothermic reaction — can be selected. Or, alternatively, a threshold for the enthalpy is given such that all reactions with an enthalpy lower than this threshold are still allowed. If in the reaction of chlorine with n-butane (Table 2) the heat of reaction is taken as the sole criterion for selection, i.e., thermodynamic product control is anticipated, then one would expect 2-chlorobutane as the main product together with a small amount of 1-chlorobutane. This is precisely what one gets. Actually the percentage of 2-chlorobutane it not as high as one would expect from the heat of reaction since the differences in the activation energies are smaller but they run parallel to the reaction enthalpies. The other products would be formed in reactions which are much less exothermic and are therefore much less likely to be obtained under normal conditions.

If one is interested in sequences of reactions, as in synthetic planning, the reaction enthalpy is applied differently for the evaluation. In this case an upper and lower bound for the energy is used. Generally, then, reactions with enthalpies around zero are to be preferred. Neither intermediates which are too low in energy should be obtained for they will not react any further, nor should they be too high in energy for they might then not be accessible because of their inherent instability.

The program for the estimation of heats of reaction is very fast; on an IBM 360/91 about 0.02 to 0.03 sec is required for calculating a reaction enthalpy. The computation times are independent of the size of the molecules since only the immediate vicinity of the bonds affected in the rearrangement is considered. A comparison with quantum mechanical methods, the alternative to our approach, is appropriate. Computation times, even with semi-empirical methods, would be many orders of magnitude higher. Furthermore, with increasing size of the molecules considered, the computation times increase very rapidly (*e.g.*, the number of integrals calculated increases with the fourth power of the number of basis orbitals for *ab initio* methods). Therefore only small systems could be handled. And finally, the values obtained generally are not yet as exact and reliable as in our approach.

6. Synthesis Tree

Applying all reaction schemes on the ensemble of molecules consisting of the target molecule and its synthesis partners with breaking of all breakable bonds generates the first level of synthetic precursors. In order to obtain sequences of synthetic reactions these precursors have now to be taken for their part as synthetic targets and must be submitted to the reaction generating process. Thus, the second level of precursors is produced. This process is repeated automatically, thereby generating a tree of synthetic precursors (Fig. 12). Branches of this tree constitute sequences of synthetic reactions. Synthetic planning is done retrosynthetically: Starting from a target molecule synthetic intermediates are produced until the compounds generated are readily accessible. Then, a branch of the tree is to be considered as brought to a close if all ends of a branch constitute available starting materials.

In EROS the synthesis tree is generated completely automatically without interference by the user to exclude him from rejecting certain reactions and favoring others just on the ground of familiarity. On the other hand, the user has various options available to tailor the development of the synthesis tree to his special needs by specifying various numbers at the start of a run. Limits on the number of levels to be generated and the numbers of reaction per molecule and per level can be given. The number of the best reactions which will be developed for the next level can also be specified. Through these options the user can decide whether he wants the program to go either more into the breadth or more into the depth of a synthesis search (see Fig. 13).

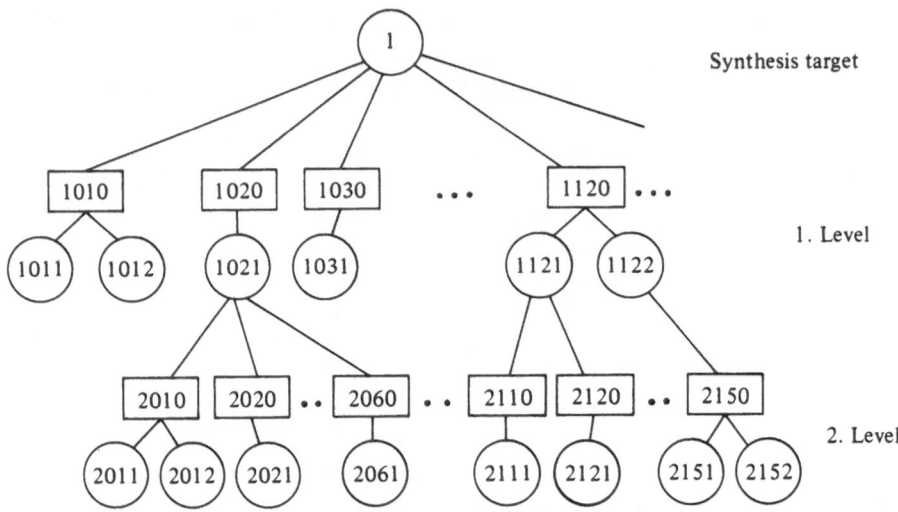

Fig. 12. Synthesis tree; ○ symbolizes reactions, □ stands for molecules. [The first digit (leftmost digit) in the numbers shows the number of the level, the next two digits the number of the synthetic reaction of this level, the last digit gives the number of the molecule of that reaction. Accordingly, 2152 refers to the second molecule in the fifteenth reaction of the second level]

To establish the end of a twig of the synthesis tree, available starting materials have to be recognized. To this end a file of commercially available compounds must be included in the program. Every molecule generated must be checked to determine whether it is contained in that list and also whether it has already been obtained previously to avoid duplicating branches in the synthesis tree. For both search operations a unique representation for molecules is required.

As atoms can be numbered in $n!$ ways, up to $n!$ different bond lists can be given for a single molecule. A program which indexes the atoms such that a unique bond list results has been developed[24]. This program also finds atoms which are constitutionally equivalent, *i.e.*, atoms which are equivalent because of global or local constitutional symmetry in the molecule.

7. Input of Molecules

The constitution of a molecule is usually specified by a constitutional formula, *i.e.*, a graph whose nodes are atoms and whose edges are bonds. It seems therefore attractive to put man-machine communication on a graphical basis, too. However, to make EROS independent of special hardware requirements, no use of graphical input devices was projected in the initial phase.

The information contained in the constitution of a molecule consists of the set of atoms and the bonds between them[25]. The input to EROS is tailored accordingly. The input routine is very flexible to allow many different forms of input; the most concise way to give the constitution of a molecule will be illustrated with an example.

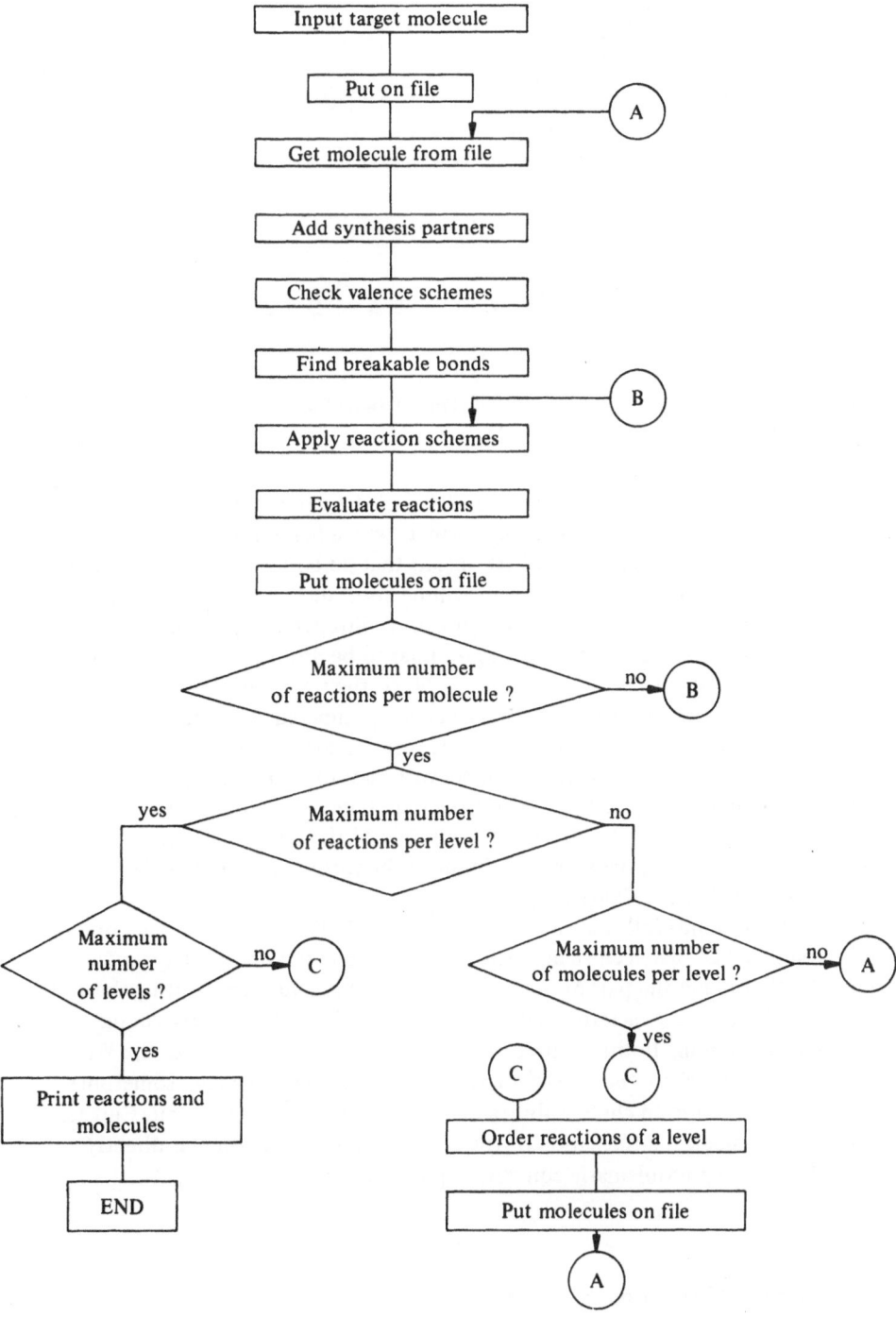

Fig. 13. Function flow for generating the synthesis tree

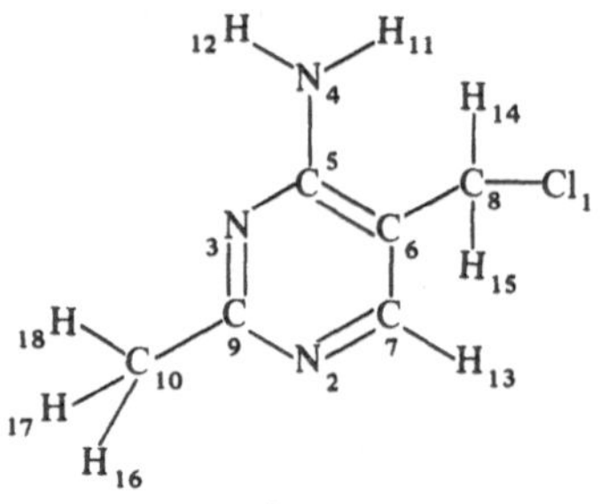

HF 1 Cl 2 N 5 C 11 H 18 H * 1 8 8 6 5 D 6 B 6 7 4 5 5 3

3 D 9 B 2 D 7 B 2 9 9 10 * *

Fig. 14. Pyrimidin-part of vitamin B_1-molecule; illustration of input

In Fig. 14 the input for the pyrimidin part of the vitamin B_1 molecule is shown. All specifications are not bound to any format; only a blank is necessary to separate individual information. The symbol HF means that no bonds to hydrogen atoms have to be given. Then, hydrogen atoms will be appended to the skeleton of the input molecule to bring carbon to a valency of 4, nitrogen to 3, etc. If other valencies are desired, *e.g.*, carbonium ions, this option has to be dropped and bonds to the hydrogen atoms have to be given explicitly. Next, the set of atoms of a molecule follows. To this end, the atoms of a molecule are numbered arbitrarily. To make the input as short as possible, atoms of the same sort are given consecutive numbers. Then, only the number of the first member of a class of atoms has to be indicated, implicitly all numbers up to the next element are taken as belonging to the same sort of atoms. For checking purposes the number of the very last atom has also to be specified. In essence, the atoms are input in the form of a structural formula, in the example of Fig. 14: $ClN_3C_6H_8$.

The list of bonds follows. Pairs of numbers symbolize bonds. Double bonds are indicated by a "D" between such a pair. If breakable bonds are to be specified on input, a "B" follows the pair of numbered atoms. This code, representing the constitution of a molecule, is rather concise. The rules for coding are very simple and require little training. In this respect our input is superior to forms like the Wiswesser Line Notation[26] where there exist a large number of rules which are sometimes quite complex and need chemically trained staff for preparing the input. Furthermore, the topological information of the constitution of a molecule is directly accessible, making problematic conversion programs unnecessary.

8. Output of Molecules

In the program, molecules are manipulated as lists of atoms and bonds. But the chemist is used to conceiving molecules through constitutional formulas. We, therefore, appended a program to EROS which generates constitutional formulas from the bond lists for output on a line printer. This program is based on concepts developed by Feldmann[27] but was extensively modified and expanded by us[28].

The restricted graphical potential of a line printer puts limitations on the quality of the constitutional formulas. The line printer cannot put symbols just anywhere on the drawing-plane but is restricted to a rectilinear grid and directions which are multiples of 45°. Furthermore, many line printers have only a small selection of symbols available. Therefore, to represent bonds, only centro-symmetric symbols are used. Figs. 15 and 16 show examples of constitutional formulas generated. If

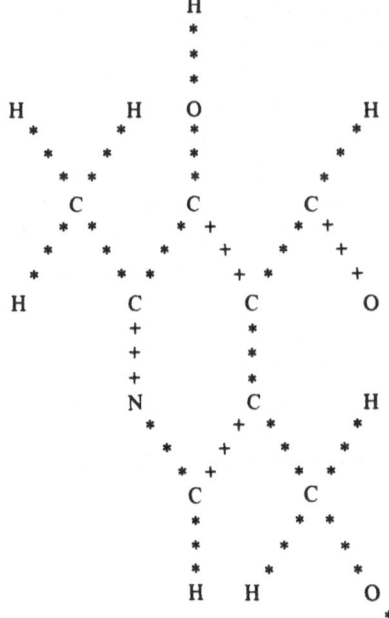

Fig. 15. Constitutional formula of pyridoxal (vitamine B_6) as generated on the line printer, an asterisk marks a single bond, a plus sign a double bond

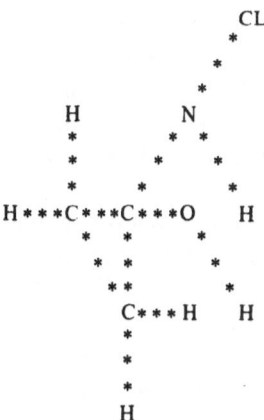

Fig. 16. Constitutional formula of N-chloro-1-amino-cyclopropanol

symbols such as the back-slash are available (not contained in the EBCDIC code!) constitutional formulas can be generated which are much more appealing to the chemist's eye. The line printer was chosen as output device for constitutional formular because this guarantees complete independence of special hardware requirements; a line printer is attached to any computer facility.

A number of options are provided for the user to adjust the constitutional formulas being output to his special desires. Atom numbers or atomic symbols, or both, can be put onto the nodes of the constitutional graph. Further, the entire constitutional formula with all hydrogen atoms, or only the skeleton of the molecule, can be drawn. In a few cases of molecules with complex polycyclic ring structures only partially complete formulas are output. Then, messages pointing to the deficiences are given which usually allow the completion of the constitutional formulas by hand.

Our structure drawing program exploits the limited potential of the line printer. Constitutional formulas of higher esthetic quality can be generated on other graphic devices, for example a graphics terminal. This amounts to some investment in the hardware of a computer facility but graphics terminals are becoming more and more accessible. We are therefore now developing a computer program which generates constitutional formulas on an interactive graphics display[29]. Using monocentric skeletons like tetrahedron, trigonal bipyramid, etc., a three-dimensional model of the molecule is generated. Going through classical potential functions of molecular mechanics this model is improved, particularly when it contains ring systems. Finally, this three-dimensional model is projected onto a plane, to yield the structure displayed on the screen. The model can be enlarged or diminished, and rotated around all three axes to allow various views into the molecule and to give information about its three-dimensional structure. Figure 17 shows an example of such a structure.

9. Technical Features

EROS is written in PL/1 and has been implemented on an IBM 360/91, an IBM 370/158, an AMDAHL 470 V6, and a Telefunken TR 440. PL/1 was chosen because it offers a number of features which make this language particularly useful for problems of this kind: Dynamic allocation of storage, operations on bit and character strings, versatile input and output operations, and extensive list and file handling capabilities. Four years of programming in PL/1 for EROS have given us no reason to regret this choice. PL/1 becomes more and more accessible; at many major computer producer a PL/1 compiler has been implemented or is being developed.

The program which generates the constitutional formulas for output on the line printer has been written in FORTRAN. It can be either called as a subroutine to the PL/1 program or, in a separate run, the lists of bonds can be taken from a file of the molecules generated previously.

The program for obtaining three-dimensional models of molecules and displaying them on a screen is written in FORTRAN on a PDP 11/45 minicomputer. The interactive graphics system is a DEC GT 44.

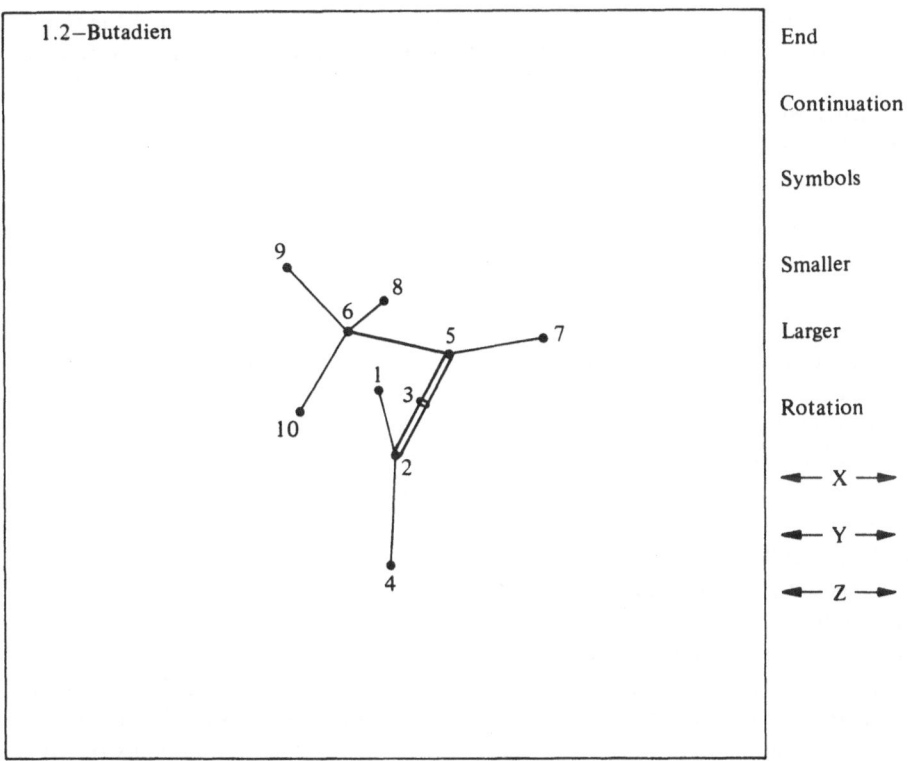

1.2—Butadien

End

Continuation

Symbols

Smaller

Larger

Rotation

← X →

← Y →

← Z →

Fig. 17. Three-dimensional model of butadiene-1.2 as generated by the structure drawing program on the display screen

EROS requires a minimum size of 300 K bytes of core storage. On an AMDAHL 470 V6, six reactions are generated per second for a molecule of 20 to 30 atoms. This includes the time for calculating the reaction energies, printing the lists of atoms and bonds, and storing and retrieving molecules and reactions on external files. Computation times increase slowly with the size of the molecules since many operations work only on the atoms directly involved in the bond rearrangement and their immediate vicinity. This makes these operation times independent of the size of the molecule. The structure program takes about 0.1 sec to generate and output a constitutional formula.

A number of options allow the user to define starting conditions for a program run to tailor them to his special problem and needs. But once the conditions are chosen, the program runs without further intervention. The user has been excluded from manipulating the course of a program run and the generation of the synthesis tree to ensure that no interesting reactions and syntheses are rejected for the sole reason that they appear unfamiliar and unusual to the chemist. As the enormous potential of the chemist as a selector to find "reasonable" syntheses is renounced, the program has to fulfill evaluation and selection, too. As a consequence, the internal evaluation has to be very carefully carried and requires excellent selection models and procedures. In an interactive program, on the other hand, uninteresting reaction paths are detected early and can be discarded immediately. To study the program's performance and the selection characteristics of EROS, an interactive version has also been developed. It gives more insight into the behaviour of chemical structures when subjected to formal reaction generating schemes and allows the recognition of basic selection principles.

10. Applications

Our main efforts have been devoted to the development and improvement of the program system EROS. Program runs were primarily done in connection with the testing of various routines. But EROS has already attained a stage where it can serve as a tool for treating problems of direct chemical interest. This has emerged from the few production runs carried out so far. The examples studied have shown the versatility and the potential of EROS as a problem solver. A small selection of problems will illustrate this.

It should be pointed out again that two fundamentally different kind of problems can be studied with EROS, depending on whether the reactions generated are taken in the forward direction or retrosynthetically. In the first case, the molecules generated from the input molecule are regarded as reaction products. Then, all compounds to be expected from the input molecules through sequences of reactions are obtained. Typical applications are: Studies on the potential use of an industrial byproduct, in which the compound is allowed to react with inexpensive and easily accessible compounds. Or, studies on the products of a compound under certain conditions, *e.g.*, in the environment where H_2O, CO_2, O_2, O_3, etc., would constitute the reaction partners. Synthetic planning belongs to the second category of problems. Here, the molecules generated from the input molecule are actually to be taken as starting materials for reactions leading to the molecules from which they were obtained in the program. The reactions proceed in a manner reversed to the way they were generated.

10.1. Valence Tautomers of Benzene

Specifying the three double bonds of a Kekulé-structure of benzene as breakable allows the generation of all valence tautomers of benzene which can be obtained by rearranging the π-bonds (Fig. 18). Breaking σ-bonds, too, leads to another valence tautomer, bicyclopropenyl. Admittedly, this problem can easily be solved with pencil and paper but this has the danger that certain forms are overlooked. For example, in a leading article on the valence tautomers of benzene[30] there was no mention of bicyclopropenyl. With multiply substituted benzenes, and in the $(CH)_8$ or $(CH)_{10}$-series, the situations are far more complex. There, it becomes very difficult to manually generate all isomers.

Employing a computer program which allows the exhaustive generation of all isomers is of great value then.

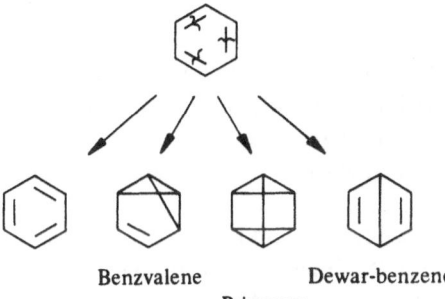

Benzvalene   Dewar-benzene
Prismane

Fig. 18. Resonance structures and valence tautomers of benzene

10.2. Isomerization and Chlorination of n-Butane

These reactions have already been discussed in Chapter 5 in connection with the calculation of heats of reaction (see Tables 1 and 2).

10.3. β-Lactam Synthesis

With this example β-lactam was chosen, a structure which is contained in the penicillin and cephalosporin antibiotics. With these compounds the synthesis of the β-lactam ring is of central importance. The target β-lactam had as synthesis partners HCl and H$_2$O. The bonds marked were taken as breakable. Figure 19 lists some of

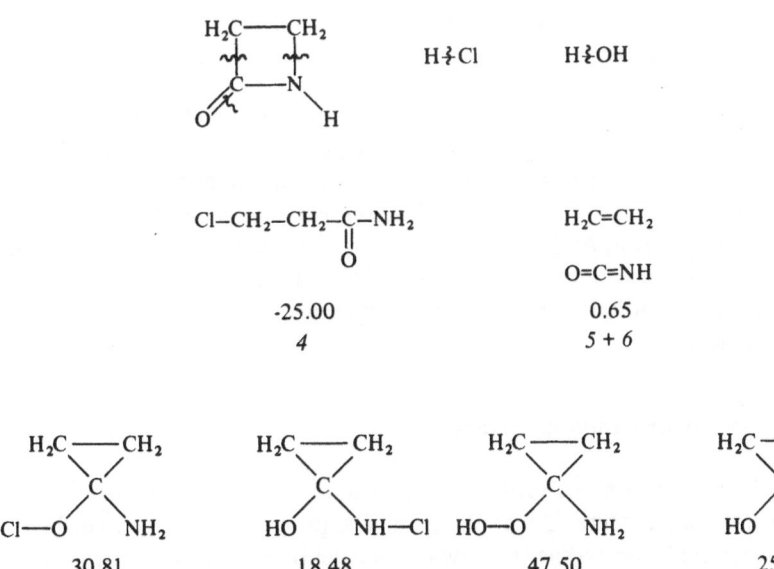

Fig. 19. β-Lactam synthesis; enthalpies in kcal/mole

the synthetic precursors thus generated. Additional synthetic reactions could be obtained if further bonds were broken. With this example, the way in which reaction enthalpies can serve as a criterion for evaluating syntheses will be discussed. In Fig. 19 values of heats of reaction are listed with the synthetic precursors. Negative values indicate that the compounds are more stable than the starting materials (*i.e.*, the synthetic target and its synthesis partner), whereas positive values indicate that the compounds are higher in enthalpy.

β-Chloropropionamide *4* is more stable than β-lactam and HCl by 25.0 kcal/mole. Therefore, it is not expected to react to these molecules as 25.0 kcal/mole would be necessary to convert it to β-lactam. The activation energy for this process should even be considerably higher than 25.0 kcal/mole. On the other hand, the reaction of olefins with isocyanates to β-lactam is estimated to be slightly exothermic by 0.7 kcal/mole. Accordingly, many syntheses of β-lactams from these precursors are known. The four compounds *7–10* of the last row in Fig. 19, however, are much higher in enthalpy and allow a look at the extreme end of the scale of exothermic reactions. All four compounds are expected to convert easily to β-lactam. But from the two isomeric pairs *7, 8* and *9, 10* the compounds *7* and *9* are so high in enthalpy that they might not be accessible. Rather, *8* and *10*, if at all, can be hoped to be attainable. Provided this is possible, they should rearrange readily to β-lactam. And, indeed, syntheses proceeding through compounds with the skeleton of *8* and *10* were discovered in 1971 by Wasserman[31].

To generalize, it might be worthwhile to look for synthetic precursors in the range of 15–30 kcal/mole if one is interested in synthetic reactions covering only a few steps. For more complex targets, where many more synthetic steps have to be anticipated, precursors which are much less high in energy have to be selected. Otherwise, after only a few steps one would end up with precursors where the energy has added up to such a high value that the compounds would no longer be accessible or stable. Rather, an enthalpy window for selecting the promising synthetic reactions has to be used with values for the heats of reaction being neither too positive nor too negative.

Returning to the synthesis of β-lactam form *8* or *10:* It should be stressed again that these synthetic reactions were generated by the program without using any prior information on them. They were generated by our formal reaction schemes and would have been among the output even if these reactions were not known at the time of the program run. Analysis of the heats of reaction as just discussed would have shown that *8* and *10* could easily convert to β-lactam. Had these reactions not been known, these considerations should have been ample reason to try to verify them in the laboratory.

10.4. Dimerization of Hexafluoropropylene Oxide

In studies for the elaboration of a synthesis of perfluoro-*n*-propyl vinyl ether[32], a synthetic precursor of structure *11* was obtained. The program suggested that this compound, *11*, would be accessible from two molecules of hexafluoropropylene oxide *12* (Fig. 20). This appeared to be quite an astonishing reaction as extensive bond reorganizations are necessary to achieve the reaction. A chemist who had to

$$F_3C-CF_2-CF_2-O-CF=CF_2$$

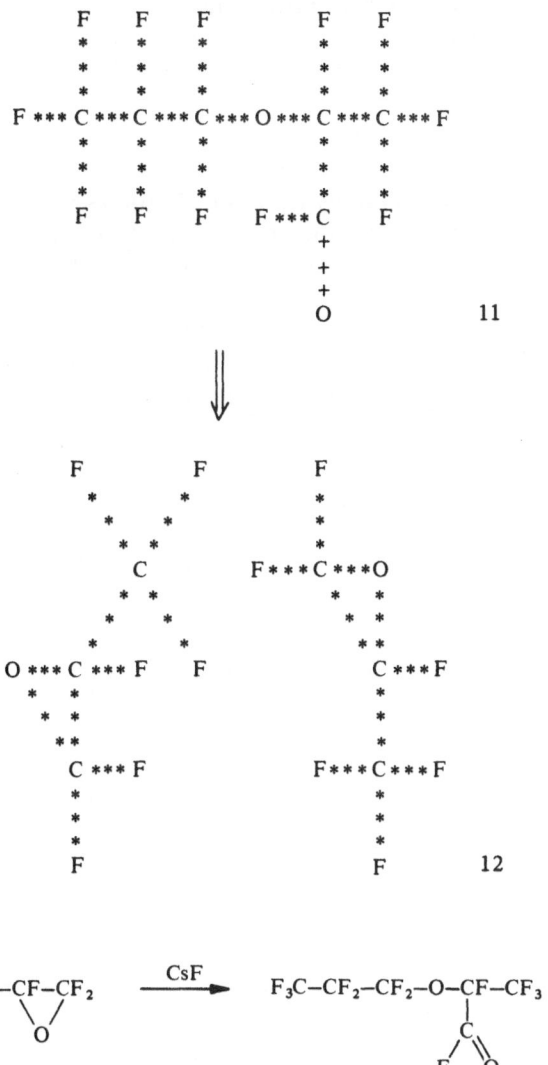

Fig. 20. Synthesis of perfluoro-n-propyl vinyl ether

evaluate that reaction and was not an expert in that field could easily be tempted to reject it as being too improbable. But this reaction does proceed. In fact it is the technical process for producing *11*, the catalyst for that reaction being cesium fluoride[33].

This example illustrates a strategy that could be taken to discover new synthetic reactions with the aid of EROS. Just the very fact of the profound simplifications in the skeleton in going from *11* to *12* should already have made this reaction attractive and given the reason to try to test its feasibility in the laboratory. In other words, the output should be scanned particularly for retro-reactions which lead to gross breakdowns of the molecular skeleton as then the corresponding synthetic reactions allow the synthesis of rather complex structures from fairly simple ones, making the reactions powerful tools in synthesis.

10.5. Synthesis of Acrylonitrile

With this example the use of EROS in studying technical processes will be illustrated. The target compound is acrylonitrile. Addition of water to the nitrile group is energetically favored when OH is bonded to the carbon atom and the hydrogen to the nitrogen atom (endothermic by 2.9 kcal/mole). But in a retrosynthetic analysis of a synthesis through a few steps, that retro-reaction should be taken which needs more energy because then the synthetic reaction itself — which is the reverse — is more exothermic. Thus the other mode of addition is to be favored. This argument holds

$$CH_2=CH-CH_3 + NH_3 + 1.5\ O_2 \longrightarrow CH_2=CH-CN + 3H_2O$$

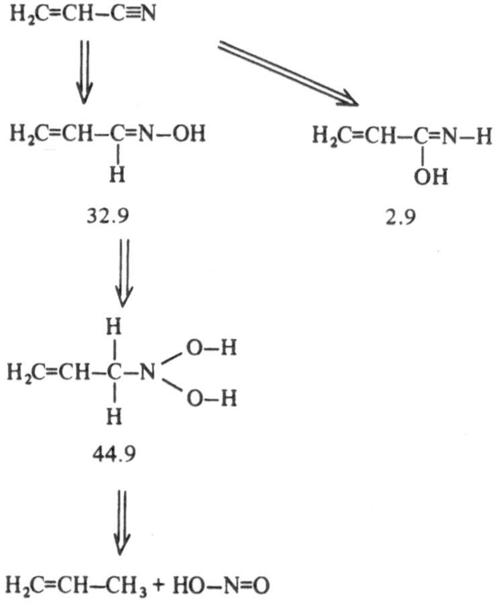

$$H_2C=CH-C\equiv N$$

$$H_2C=CH-\underset{\underset{H}{|}}{C}=N-OH \qquad\qquad H_2C=CH-\underset{\underset{OH}{|}}{C}=N-H$$

$$32.9 \qquad\qquad\qquad\qquad 2.9$$

$$H_2C=CH-\underset{\underset{H}{\overset{H}{|}}}{C}-N\!\!<^{O-H}_{O-H}$$

$$44.9$$

$$H_2C=CH-CH_3 + HO-N=O$$

$$-20.0 \qquad\qquad\qquad \text{Fig. 21. SOHIO-process; energies in kcal/mole}$$

for the addition of a second molecule of water; the more, as in the following step the sign of the reaction enthalpy reverses and brings the overall reaction enthalpy more into balance. With this last step, propene and nitrous acid were obtained. The study was ended as it was realized that nitrous acid can be generated by oxidation of ammonia. Thus, the entire SOHIO-process of making acrylonitrile on an industrial scale was obtained. Only later was it recognized that the analysis had generated a second industrial process for producing acrylonitrile; *i.e.*, the Dupont process which works with propene and nitrous acid directly. Both processes must appear quite unusual for chemists trained in thinking of "normal" synthetic reactions and it will therefore take a while until reactions of this kind are incorporated into computer programs working with libraries of synthetic reactions. In our approach, however, these industrial processes were generated quite naturally; no special provisions were necessary to obtain them. The planning of industrial syntheses and of syntheses in the laboratory is not of a fundamental formal difference in our approach. It can be done with the same program system, the difference being only in the accessible conditions (*e.g.*, reaction enthalpies).

10.6. Biosynthesis of Pyridoxal

As the biosynthesis of pyridoxal (vitamin B_6) is not yet completely understood, this example was chosen to demonstrate the behaviour of EROS in treating biochemical problems. As shown in Fig. 22, ammonia, glyceraldehyde, and a pentose were obtained from pyridoxal and water through several reaction steps. As the ques-

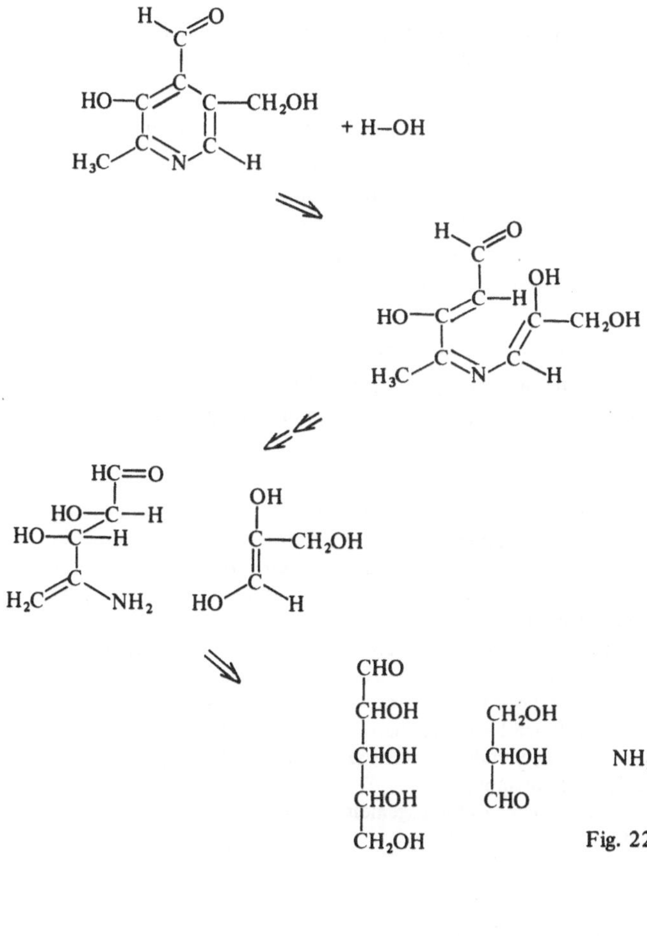

Fig. 22. Biosynthesis of pyridoxal?

123

tion mark indicates, this study does not imply that the biosynthesis of pyridoxal necessarily proceeds along those paths. But it is a hypothesis which seems attractive enough to warrant experimental examinations. Indeed, isotopic labelling experiments indicate that at least one C-3 fragment of pyridoxal is provided by glyceraldehyde-3-phosphate[34]. Further, it appears that the other C-atoms, too, are derived from the carbohydrate metabolism.

11. Conclusion

It has been demonstrated that a mathematical model of constitutional chemistry can serve as a basis for a computer program to automatically generate sequences of reactions and for planning organic syntheses. This gives credibility to other applications of this model.

A finite — and indeed a rather small — amount of computer time and storage is required to generate a tree of reactions; the danger of an information explosion has been contained. Due to the formal character of the reaction generating process all conceivable reactions are obtained. This open-ended approach permits the discovery of completely novel synthetic reactions.

The field is now open for studying a great variety of problems of immediate interest to the chemist with the aid of a computer. Problems can range from the planning of organic syntheses and studying syntheses of industrial and biochemical interest to the generation of the products of a compound in a certain environment.

In the design of the system, care was taken to achieve a versatile and flexible program to allow adaption of the program run to the problem given. On the other hand, excluding the user from interfering during a run allows full advantage to be taken of the innovativeness of the system and the study of its artificial intelligence capabilities. In developing several versions of the program over the years we have seen the educability of our system.

The design of selection and evaluation procedures gives new insights into the driving forces of chemical reactivity and the principles of synthetic planning. This appears to us as being the most challenging and exciting part of developing such a computer program.

Acknowledgments. The work reported here has roots in the efforts of many people. Decisive were the insights gained with our prototype program CICLOPS. During that phase, numerous discussions with Professor Ivar Ugi and Dr. Paul Gillespie laid the foundations to our approach. The dedication of Janet Blair and Carol Gillespie brought CICLOPS to reality. Their excellent programming style was a model for our work.

The continuous encouragement by Professor Ugi is greatly appreciated. Able students — Oliver Dammer, Erika Diener-Weselsky, Peter Jacob, Manfred Oswald, and Josef Thoma — worked on various programming projects; our thanks is extended to them. The structure drawing program is based on concepts developed by Dr. Richard Feldmann of NIH and was extensively modified and expanded here by Wolfgang Schubert, and Josef Thoma. Helpful discussions with Josef Friedrich and Wolfgang Schubert are appreciated.

The financial support of the Deutsche Forschungsgemeinschaft and of Stiftung Volkswagenwerk is gratefully acknowledged.

Note Added in Proof

A recent publication (Hill, R. E., Miura, I., Spenser, I. D., J. Amer. Chem. Soc. *99,* 4179 (1977)) shows that five carbon atoms of pyridoxol are derived from $[1,3 - {}^{13}C_2]$ glycerol, i.e., that three molecules of glycerol are incorporated into pyridoxol. This gives additional support to our scheme.

12. References

[1] Corey, E. J., Orf, H. W., Pensak, D. A.: J. Amer. Chem. Soc. *98,* 210 (1976), and earlier references

[2] Wipke, W. T., Gund, P.: J. Amer. Chem. Soc. *98,* 8107 (1976), and preceding papers

[3] Gelernter, H., Sridharan, N. S., Hart, H. J., Yen, S. C., Fowler, F. W., Shue, H. J.: Topics Curr. Chem. *41,* 113 (1973)

[4] Dugundji, J., Ugi, I.: Topics Curr. Chem. *39,* 19 (1973)

[5] Ugi, I., Gillespie, P.: Angew. Chem. *83,* 980, 982 (1971); ibid. Internat. Edit. *10,* 914, 915 (1971)
Ugi, I., Gillespie, P. D., Gillespie, C.: Trans. N. Y. Acad. Sci. *34,* 416 (1972)

[6] Brandt, J., Friedrich, J., Gasteiger, J., Jochum, C., Schubert, W., Ugi, I.: Angew. Chem., in preparation

[7] Brandt, J., Friedrich, J., Gasteiger, J., Jochum, C., Schubert, W., Ugi, I.: Proc. ACS-Meeting, New York, April 1976

[8] Brandt, J., Friedrich, J., Gasteiger, J., Jochum, C., Schubert, W., Ugi, I.: Proc. III. Intern. Conf. on Computers in Chemical Research, Education, and Technology, Caracas, Venezuela, July 1976

[9] Blair, J., Gasteiger, J., Gillespie, C., Gillespie, P., Ugi, I., in: Computer representation and manipulation of chemical information (ed. W. T. Wipke, S. Heller, R. Feldmann, E. Hyde, Wiley). New York 1974

[10] Blair, J., Gasteiger, J., Gillespie, C., Gillespie, P. D., Ugi, I.: Tetrahedron *30,* 1845 (1974)

[11] Ugi, I., Gasteiger, J., Brandt, J., Brunnert, J. F., Schubert, W.: IBM-Nachr. *24,* 185 (1974)
Ugi, I.: IBM-Nachr. *24,* 180 (1974)

[12] Review Articles:
Thakkar, A. J.: Topics Curr. Chem. *39,* 3 (1973)
Bersohn, M., Esack, A.: Chem. Rev. *76,* 269 (1976)

[13] Bersohn, M.: Bull. Chem. Soc. Japan, *45,* 1897 (1972)
Esack, A., Bersohn, M.: J. C. S. Perkin I, 2463 (1974)

[14] Barone, R., Chanon, M., Metzger, J.: Rev. Inst. Fr. Pet. *28,* 771 (1973)

[15] Blower, P. E., Jr., Whitlock, H. W., Jr.: J. Amer. Chem. Soc. *98,* 1499 (1976)

[16] Weise, A.: Z. Chem. *15,* 333 (1975)

[17] Yoneda, Y.: private communication

[18] Bart, J. C. J., Garagnani, E.: Z. Naturforsch. *31b,* 1646 (1976)

[19] Gasteiger, J.: in preparation

[20] Cox, J. D., Pilcher, G.: Thermochemistry of organic and organometallic compounds. London: Academic Press 1970

[21] Benson, S. W.: Thermochemical kinetics. Second Edit. New York: Wiley 1976
S. W. Benson, *et al.:* Chem. Rev. *69,* 279 (1969)

[22] a) Allen, T. L.: J. Chem. Phys. *31,* 1039 (1959)
b) Kalb, A. J., Chung, A. L. H., Allen, T. L.: J. Amer. Chem. Soc. *88,* 2938 (1966)

[23] Pihlaja, Kankare, J.: Acta Chem. Scand. *23,* 1745 (1969)
Pihlaja, K.: Acta Chem. Scand. *25,* 451 (1971)
Pihlaja, K., Tuomi, M. L.: Acta Chem. Scand. *25,* 465 (1971)

[24] Jochum, C., Gasteiger, J.: J. Chem. Inf. Comput. Sci. *17,* 113 (1977)

J. Gasteiger and C. Jochum

25) Lynch, M. F., Harrison, J. M., Town, W. G., Ash, J. E.: Computer handling of chemical structure information. London: Macdonald 1971
26) Smith, E. J.: The Wiswesser line-formula chemical notation. New York: McGraw-Hill 1968
27) Feldmann, R. J., in: Computer representation and manipulation of chemical information. (eds. W. T. Wipke, S. R. Heller, R. J. Feldmann, E. Hyde). London: Wiley 1974, p. 55
28) Feldmann, R. J., Gasteiger, J., Schubert, W., Thoma, J.: in preparation
29) Gasteiger, J., Schubert, W., Thoma, J.: unpublished results
30) Viehe, H. G.: Angew. Chem. *77*, 768 (1965)
31) Wasserman, H. H., Adickes, H. W., deOchoa, O. E.: J. Amer. Chem. Soc. *93*, 5586 (1971)
32) We thank Dr. H. Nickelsen, Hoechst AG, for drawing our attention to that problem
33) Moore, E. P., Jr., U. S. Pat. 3.322.826 (1967)
34) Hill, R. E., Spenser, I. D.: Can. J. Biochem. *51*, 1412 (1973)
Review:
Plaut, G. W. E., Smith, C. M., Alworth, W. L.: Ann. Rev. Biochem. *43*, 899 (1974)
We thank Dr. Palm, University of Würzburg, for making us aware of that reference

Received, July 11, 1977

Author Index Volumes 26-74

Reactivity and Structure

Concepts in Organic Chemistry

Editors: K. Hafner, J.-M. Lehn, C. W. Rees,
P. v. Ragué Schleyer, B. M. Trost, R. Zahradník

Volume 4
W. P. Weber, G. W. Gokel

Phase Transfer Catalysis in Organic Synthesis

100 tables. XV, 280 pages. 1977
ISBN 3-540-08377-4

Contents: Introduction and Principles. – The Reaction of Dichlorocarbene with Olefins. – Reactions of Dichlorocarbene with Non-Olefinic Substrates. – Dibromocarbene and Other Carbenes. – Synthesis of Ethers. – Synthesis of Esters. – Reactions of Cyanide Ion. – Reactions of Superoxide Ions. – Reactions of Other Nucleophiles. – Alkylation Reactions. – Oxidation Reactions. – Reduction Techniques. – Preparation and Reactions of Sulfur-containing Substrates. – Ylids. – Altered Reactivity. – Addendum: Recent Developments in Phase Transfer Catalysis.

Volume 5
N. D. Epiotis

Theory of Organic Reactions

69 figures, 47 tables. XIV, 290 pages. 1978
ISBN 3-540-08551-3

Contents: One Determinental Theory of Chemical Reactivity. – Configuration Interaction Overview of Chemical Reactivity. The Dynamic Linear Combination of Fragment Configurations Method. – Even-Even Intermolecular Multicentric Reactions. – The Problem of Correlation Imposed Barriers. – Reactivity Trends of Thermal Cycloadditions. – Reactivity Trends of Singlet Photochemical Cycloadditions. – Miscellaneous Intermolecular Multicentric Reactions. – $\pi + \sigma$ Addition Reactions. – Even-Odd Multicentric Intermolecular Reactions. – Potential Energy Surfaces for Odd-Odd Multicentric Intermolecular Reactions. – Even-Even Intermolecular Bicentric Reactions. – Even-Odd Intermolecular Bicentric Reactions. – Odd-Odd Intermolecular Bicentric Reactions. Potential Energy Surfaces for Geometric Isomerization and Radical Combination. – Odd-Odd Intramolecular Multicentric Reactions. – Even-Even Intramolecular Multicentric Reactions. – Mechanisms of Electrocyclic Reactions. – Triplet Reactivity. – Photophysical Processes. – The Importance of Low Lying Nonvalence Orbitals. – Divertissements. – A Contrast of "Accepted" Concepts of Organic Reactivity and the Present Work. – Author Index. – Subject Index.

Volume 6
M. L. Bender, K. Komiyama

Cyclodextrin Chemistry

14 figures, 37 tables. X, 96 pages. 1978
ISBN 3-540-08577-7

Contents: Properties. – Inclusion Complex Formation. – Catalyses by Cyclodextrins Leading to Practical Usages of Cyclodextrins. – Covalent Catalyses. – Noncovalent Catalyses. – Asymmetric Catalyses by Cyclodextrins. – Improvement by Covalent and Noncovalent Modification.

Springer-Verlag
Berlin
Heidelberg
New York

Topics in Current Chemistry

Springer-Verlag
Berlin
Heidelberg
New York